NEW JERSEY ASK8

SCIENCE TEST

JACKIE HALAW
Science Teacher
Crossroads South Middle School
South Brunswick, NJ

All inquiries should be addressed to:
Barron's Educational Series
250 Wireless Boulevard
Hauppauge, NY 11788
www.barronseduc.com

Library of Congress Control Number: 2008024291

ISBN-13: 978-0-7641-4028-0
ISBN-10: 0-7641-4028-0

Library of Congress Cataloging-in-Publication Data

Halaw, Jackie.
New Jersey ASK grade 8 science test / Jackie Halaw.
p. cm.
Includes index.
ISBN-13: 978-0-7641-4028-0
ISBN-10: 0-7641-4028-0
1. Science—Study and teaching (Elementary)—New Jersey (States) 2. Science—Examinations—Study guides. 3. New Jersey Assessment of Skills and Knowledge—Study guides. 4. Eighth grade (Education)—New Jersey (State) I. Title. II. Title: New Jersey ASK grade eight science test.

LB1585.3.H345 2008
372.35'044—dc22 2008024291

PRINTED IN THE UNITED STATES OF AMERICA

9 8 7 6 5 4 3

CONTENTS

IMPORTANT NOTE

Please consult *http://www.state.nj.us/education/assessment* for all the latest information on Grade 8 Science New Jersey Assessment of Skills and Knowledge (NJ ASK).

INTRODUCTION

Thank you for selecting this test preparation guide. This book is designed to prepare New Jersey middle school students for the science portion of the New Jersey Assessment of Skills and Knowledge administered in eighth grade (NJ ASK8). It offers general science content review, test-taking strategies, dozens of practice questions, and two full-length practice tests. Answers and explanations are provided for all of the multiple-choice and open-ended questions found in this test guide.

This guide will acquaint students with the format of the test and the content that it assesses. Students will build confidence and prepare for success by learning standardized test-taking strategies and by reviewing the content supporting the New Jersey Core Curriculum Content Standards and the NJ ASK8 science test specifications.

TO THE STUDENT

This book has been written to help you prepare to take the science portion of the NJ ASK8. Although the NJ ASK8 is given toward the end of eighth grade, it will assess your understanding of all of the science you have learned since fourth grade. This may seem overwhelming, but if you set aside some time to review using this book, you will feel confident and ready on the day of the assessment.

This book is certainly not a substitute for science learning in the classroom; it is a supplement to the hands-on, inquiry-based science lessons in your school. As you read through this book, try to make connections by recalling science lessons, activities, experiments, and investigations you performed at school. Pay close attention to the objectives in the beginning of each section and the key terms in boldface through the book. Use the objectives to stay focused as you read. Do not think of the key terms as vocabulary terms to memorize; consider them to be concepts that you must understand. You should be comfortable with using these science key terms in conversation and in written explanations of science activities. Be sure to read closely the sections on test preparation strategies later in the preface, and good luck on the NJ ASK8!

NOTE TO PARENTS

Thank you for selecting this assessment preparation guide for your child. The science portion of NJ ASK8 is designed to provide information to your school about how well your child is achieving in the science content area. The NJ ASK8 is an assessment based on the New Jersey Core Curriculum Content Standards. Your school will use the scores to help plan for the educational needs of its students.

You can help your child prepare for the NJ ASK8 by having your child set aside some time to review using this book. Make sure that he or she is well rested and nourished for school, especially on assessment days. Encourage your child to complete his or her homework in a quiet, consistent location each day. Visit the local library and borrow books that support and extend the learning that is taking place at school in their science classroom. Visit local museums and science centers, and ask your child to communicate what he or she sees and experiences. Ask "why?" and "how?" questions. Above all else, demonstrate the value of education for your child by being a lifelong learner yourself. Thank you again for selecting this guide for your child.

NOTE TO TEACHERS

This book can be used to help prepare your students for the science portion of the NJ ASK8. Students must learn science by doing science; hence, this book should be considered only as a supplement to the inquiry-based, hands-on learning in your student-centered classroom. Many teachers worry about state standards tests, seeing them more as an assessment of their teaching rather than of student learning. If that is the case, consider the following strategies as you prepare your students for the science portion of the NJ ASK8:

1. Be aware of the New Jersey Core Curriculum Content Standards (NJCCCS) for science, NJ ASK8 Test Specifications, Science Standards Clarification documents, and released sample test items from the New Jersey Department of Education. All of your lessons should be correlated to the NJCCCS for science.

The New Jersey Department of Education web site has published all of the documents mentioned above online. They are invaluable in the curriculum and lesson planning process. Utilize these resources along with their bank of lesson activities and projects.

2. There is a list of key terms at the end of each chapter in this book. Definitions of these terms appear in the Glossary at the back of the book. Encourage students to think of these terms as concepts to understand, not vocabulary words to memorize. After the class reads each section, post these terms in your classroom, perhaps on a word wall. Students should be able to explain these terms (concepts) and use this science terminology as they communicate ideas both verbally and in written form.
3. Mirror your classroom tests and quizzes to match the format of the science portion of NJ ASK8. Or, consider including a NJ ASK8 section on your assessments with sample test questions relating to your topic of study. Notice that none of the multiple-choice questions have "all of the above" or "none of the above" as possible responses. By formatting your assessments in a similar way, your students will gain confidence.
4. The multiple-choice questions on the NJ ASK8 require higher levels of thinking. Students are allotted one minute to read, consider, and select a response. Many students give up too quickly. Encourage your students to think about the questions and process the possible choices. When posing questions to your students, give them "wait time" to model good thinking strategies. Then, through Think-Pair-Share, allow students to pair up and share their responses. This allows all students to be accountable for a response and gives them the opportunity to communicate their ideas.

WHAT IS ON THE SCIENCE PORTION OF THE NJ ASK8?

The science assessment consists of four sections. Each section includes fifteen multiple-choice questions and one open-ended question. In general, students will have twenty minutes to complete each section. After the twenty minutes have passed, students are not permitted to go back and answer questions in that section. One minute is allotted for each multiple-choice question, and five minutes are allotted for each of the open-ended questions. You are allowed to respond to the open-ended question in paragraph form and with tables and diagrams. Even though you are not penalized for spelling or grammar, try to be clear and detailed in your responses.

The questions on the NJ ASK8 will assess students' ability to make interpretations, apply their knowledge, and formulate explanations. It is an assessment of the ability of students to apply concepts and science process skills. The test questions are mixed within each section, but overall life science questions compose 40% of the test, physical science questions compose 30% of the test, and earth science questions compose the remaining 30% of the assessment.

The NJ ASK8 is based on the New Jersey Core Curriculum Content Standards for science. The content within this book reflects the knowledge and skills an eighth grade student should possess relating to each of these standards.

N.J. Standard 5.1—Scientific Processes
All students will develop problem-solving, decision-making, and inquiry skills, reflected by formulating usable questions and hypotheses, planning experiments, conducting systematic observations, interpreting and analyzing data, drawing conclusions, and communicating results.

N.J. Standard 5.2—Science and Society
All students will develop an understanding of how people of various cultures have contributed to the advancement of science and technology, and how major discoveries and events have advanced science and technology.

N.J. Standard 5.3—Mathematical Applications
All students will integrate mathematics as a tool for problem solving in science, and as a means of expressing and/or modeling scientific theories.

N.J. Standard 5.4—Nature and Process of Technology
All students will understand the interrelationships between science and technology and develop a conceptual understanding of the nature and process of technology.

N.J. Standard 5.5—Characteristics of Life
All students will gain an understanding of the structure, characteristics, and basic needs of organisms and will investigate the diversity of life.

N.J. Standard 5.6—Chemistry
All students will gain an understanding of the structure and behavior of matter.

N.J. Standard 5.7—Physics
All students will gain an understanding of the natural laws as they apply to motion, forces, and energy transformations.

N.J. Standard 5.8—Earth Science
All students will gain an understanding of the structure, dynamics, and geophysical systems of the Earth.

N.J. Standard 5.9—Astronomy and Space Science
All students will gain an understanding of the origin, evolution, and structure of the universe.

N.J. Standard 5.10—Environmental Studies
All students will develop an understanding of the environment as a system of interdependent components affected by human activity and natural phenomena.

TEST-TAKING STRATEGIES

Here are a few test-taking strategies to follow for the NJ ASK. Consider them now, and read them through again the day before the test.

1. Eat a nutritious breakfast, but don't overdue it. You don't want a stomachache during the test.
2. Get plenty of sleep.
3. Listen to and read all directions.
4. Budget your time wisely. Remember, you are given one minute to complete each multiple-choice question and five minutes for each open-ended question. Many students need more time for the open-ended questions. Consider this as you work.
5. Don't spend too long on any one multiple-choice question. If you are still stuck after two or three minutes, take a guess and come back to it if you have time later.
6. Fill in your answer sheet carefully. If you skip a question or enter the answer in the wrong space, you may have to spend valuable time fixing your answer sheet rather than thinking about the questions. Be careful.
7. If you have extra time, go back and check your answers.

HOW TO USE THIS TEST GUIDE

There are many ways to use this book to help prepare you for the NJ ASK. The first step you may want to take is to simply survey the book; look at the table of contents, the format of each section, the practice test questions, and the glossary. You will notice that there are five chapters: The Nature of Science, Characteristics of Life, Environmental Studies, Physical Science, and Earth and Space Science. Each chapter is divided into three or four sections. In the beginning of each section, there are objectives. Key concepts and important scientific terms are indicated within each section. Within each section, there are examples (practice questions) that relate to the topic on that particular page. Additionally, a practice quiz is located at the end of each chapter. There are two full-length practice tests, a glossary, and an index in the back of the book.

You may choose to begin by simply reading from the first chapter and moving through the book sequentially. Alternatively, if you know that you need more review in certain areas, you may choose to skip around. Another method of using this book is to take either the full-length practice tests or chapter quizzes first. You can then diagnose or determine which areas you should focus on more heavily in your review.

Chapter 1

THE NATURE OF SCIENCE

Science is more than just a bunch of facts to memorize; it is a way of thinking about the natural world, and it is a method of solving problems. Scientists investigate these problems using a variety of processes like careful observation, classification, and data analysis. In this chapter, we will explore the methods scientists use to inquire about our world, and we will examine the role of science and technology in our society. Furthermore, we will see how scientists use mathematics to communicate ideas, make measurements, and analyze data. This chapter is divided into three sections:

- The Processes of Science
  - Science as Inquiry
  - Designing an Experiment
  - Collecting, Organizing, and Interpreting Data
  - Making Inferences and Drawing Conclusions
  - Safety
- Science, Society, and Technology
  - The Role of the Scientific Community in Our Society
  - Science and Technology
  - Great Technological Achievements in Science
  - Lives and Contributions of Important Scientists
- Mathematical Applications
  - Measurement
  - Analyzing Data in Tables and Graphs

NOTE TO STUDENTS

Thoroughly read each section. As you read, pay close attention to the key concepts and try to answer the example questions found in each section. Answers to examples in this chapter are on page 28. At the end of the Chapter 1, assess your knowledge of the nature of science with the practice quiz. Then, check your answers using the answer key provided on page 235.

THE PROCESSES OF SCIENCE

After reading this section you should be able to:

- Distinguish between qualitative and quantitative observations.

- Identify questions and make predictions that can be tested through scientific investigations.
- Design a controlled experiment.
- Collect, organize, and analyze data from investigations.
- Understand safety procedures that should be followed during scientific investigations.

Science is the study of the physical world and its **phenomena**, or observable natural events. Scientists are very careful and deliberate as they design and conduct investigations of our natural world. As they study our world, they are curious and open-minded, but they are also skeptical of information they uncover. Scientists always wonder why things happen the way they do, and they are always asking new questions. In this section, we will explore the processes scientists use to investigate the physical world.

SCIENCE AS INQUIRY

Inquiry is the act of questioning or investigating. Scientists use the processes of inquiry to question and investigate the natural world. Scientific inquiry is the way that we develop explanations about the physical world through observing, testing, experimenting, and thinking about our investigations. Through observations and investigations, we have gained an enormous amount of the scientific knowledge about the universe we live in. This knowledge is always changing as we discover new information through improved techniques and new technologies.

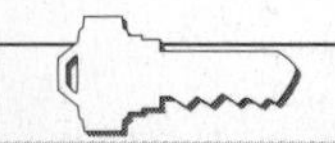

KEY CONCEPT

Instruments, like telescopes, thermometers, and microscopes, are used to help observe our world and collect data.

As we strive to understand the world, we make many observations directly with our senses. We see the Sun appear to rise and set each day. We touch the sand on the beach and notice many different-sized grains. A scientific **observation** is noticing an event or specimen in a careful and thorough way and then recording what is observed in a detailed manner. Observations are also called **data**. In addition to our human senses, we also use technological **instruments**, like telescopes, thermometers, and microscopes, to help observe our world and collect data. These instruments help us observe phenomena that we cannot directly observe with our senses, and they help us make measurements.

There are two types of observations, **qualitative** and **quantitative**. A qualitative observation describes. For example, "the worm moved from the dry soil to the damp soil" and "the liquid changed from dark brown to red" are qualitative observations. A quantitative observation is a measurement. For example, "the length of the piece of wood is two feet" and "the temperature at 3:00 P.M. was 60 degrees Fahrenheit" are quantitative observations. Scientists must make careful and detailed observations and, whenever possible, use measurements, which can be checked and repeated. Quantitative observations are usually preferred because they are more precise and because they enable scientists to easily compare their observations with others.

Our observations about the world around us usually lead to new questions. The first step of science inquiry occurs when we realize we do not know something. Our curiosity drives us to explore, investigate, and try to answer questions or solve problems. These questions can be scientific or nonscientific. A scientific question is one that can be tested through scientific investigations. For example, the question "how does fertilizer affect the growth of bean plants?" is a scientific question. If the question focuses on personal preferences, morals and values, or the supernatural, then it is not scientific. For example, the question "is hip-hop music better than country music?" is nonscientific because it deals with personal preferences. Scientific questions can be studied and answered by collecting and analyzing data and developing explanations based on data and evidence.

Example 1

Which of the following problems can be best tested through a scientific investigation?

A. Why don't some cats like water?

B. Are cats better pets than dogs?

C. Is vegetarianism better than eating meat?

D. How does the temperature of water affect how much sugar can dissolve in it?

See page 28 for the answer.

Once a question is selected, then an investigation is designed to help answer that question. Through the investigation, observations are made, and data is collected. The evidence gathered helps the scientist formulate a conclusion or answer the original question. It is then that the scientist communicates the process of the investigation and the results. Very often, this process leads to further questions that can be investigated.

Different types of scientific questions require different kinds of investigations. Some investigations might involve observing organisms, whereas others might require collecting specimens. If a phenomenon cannot be directly observed, then an investigation might require a model to be built. For example, since the Earth's interior or the Earth-Moon-Sun system cannot be observed directly, a model can be drawn or built that represents this phenomenon.

DESIGNING AN EXPERIMENT

Some questions require a scientist to design an experiment. When observing and investigating phenomena, scientists look for cause-and-effect relationships. Through experiments, scientists test how changing one item causes something else to change in a predictable way. These changing items are called variables. A **variable** is any part, or component, of an experiment that can change or be changed. For example, if we were investigating the problem, "how does the amount of fertilizer affect the growth of bean plants?" then some of the variables would be the type of fertilizer, the amount of fertilizer, the type of soil, and the height of the bean plant.

When designing an experiment, you only want to change, or test, one variable at a time. All of the other variables must stay constant, or controlled. For example, if we were testing "how does the amount of fertilizer affect the growth of bean plants?" then we would only change the amount of fertilizer given to each of the bean plants. This variable is called the **independent variable**, or **manipulated variable**. All of the other variables must stay the same; all of the plants must be given the same type of fertilizer and the same amount of soil and water. We call the variables that stay the same **controlled variables**, or **constants**. If you were to test more than one variable during an experiment, then you would not be sure which variable caused the change. In our experiment, the growth of the bean plants will change in response to the amount of fertilizer the plant receives. The variable that changes in response to the independent variable is called the **dependent variable**, or **responding variable**. So in this experiment, the growth of the bean plant is called the dependent or responding variable.

> **KEY CONCEPT**
>
> When designing an experiment only one variable can be purposely changed; all other variables must stay constant.

Here is another example. A student wanted to investigate how the temperature of water affects how much sugar dissolves in water. The student placed five cups on a table. Each cup was the same size, and she put the same amount of water and sugar into each. The water in each of the five cups was 50, 60, 70, 80, and 90 degrees

Fahrenheit. Each cup was stirred the same length of time and at the same speed. After ten minutes, she observed each cup and noted the amount of sugar that remained. In this experiment, the independent (or manipulated) variable was the temperature of the water, and the dependent (or responding) variable was the amount of sugar dissolved. The controlled variables (or constants) were the amount of water, the size of each cup, the amount of sugar, and the stirring time and speed.

Example 2

A team of students investigated how the amount of moisture affects the growth of plants. They set up five bean plants in a windowsill. Each plant was in the same type of container, had the same type of soil, and received the same amount of sunlight each day. Each plant was watered with a different amount of water each day. Plant #1 was watered with 100 mL of water each day. Plant #2 received 150 mL of water, plant #3 received 200 mL of water, plant #4 received 250 mL of water, and plant #5 received 300 mL of water each day. Which component of the experiment was the independent variable?

A. the growth of the plants

B. the type of container

C. the type of soil

D. the amount of water it received

See page 28 for the answer.

After you decide which variable you would like to test, then you will create a preliminary explanation, or hypothesis. The **hypothesis** is a guess based on previous observations or prior knowledge, and this hypothesis will be tested during the investigation. The hypothesis predicts how the manipulated variable will affect the responding variable in your experiment. A hypothesis is often written in the form of an "if . . . then . . . " statement. For example, the statement "if ice cubes are smaller, then they will melt faster" is a hypothesis. In this experiment, the size of the ice cubes is the manipulated variable, and the melting time is the responding variable. Even if a hypothesis is shown not to be true through an experiment, it is still valuable because it can lead to new discoveries or other questions that can be investigated.

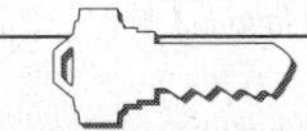

KEY CONCEPT

The hypothesis predicts how the manipulated variable will affect the responding variable in your experiment.

Example 3

Which of the following is an example of a hypothesis?

A. The boy is six feet tall.

B. Why is the sky blue?

C. If the height of a diver increases, then he will make a bigger splash in the water.

D. The water was clear, and when the powder was added, the water turned green.

See page 28 for the answer.

Variables are important to understand when designing an experiment because you want to conduct a controlled experiment. In a **controlled experiment**, only one variable is tested at a time. It is also wise to include a control. A **control** is a parallel experiment in which no variables have been manipulated, or changed. For example, if you were testing "how does the amount of fertilizer affect the growth of bean plants?" it would be wise to include one plant that is not given any fertilizer at all. This plant could be used for comparison against the plants that were given fertilizer. This would help the investigator be absolutely sure that the independent variable, fertilizer in this case, really does affect the dependent variable, the growth of the bean plant. To design a fair experiment, it is also a good idea to perform several tests. This is called doing **repeated trials**. By replicating the experiment a few times through repeated trials, you will reduce the effects of errors and unidentified variables on your data and results.

Example 4

A student wanted to test the hypothesis, if salt is added to water, then the freezing point of the water will be lower. The student recorded that pure water froze at 32 degrees Fahrenheit, while the saltwater froze at 20 degrees Fahrenheit. After collecting this data, the student wanted to confirm or check the results. The student should repeat the experiment using

A. exactly the same setup as the original experiment.

B. a different amount of water.

C. a different amount of salt.

D. a different shaped container.

See page 28 for the answer.

COLLECTING, ORGANIZING, AND INTERPRETING DATA

Before conducting an experiment, it is important to determine how the data and observations will be collected and recorded. Data needs to be recorded in a systematic, organized way. By organizing data in a data table or chart, scientists can easily record and interpret the results. When creating a data table or chart, the manipulated or independent variable is place in the left column and the responding variable is placed in the right column.

Independent Variable	Dependent Variable

After the data is recorded, a graph is usually created. A graph converts the data into a visual image, which is often easier to interpret and analyze than a data table. There are two types of graphs that are commonly used to organize and analyze data, bar graphs, and line graphs. Descriptive or qualitative data is usually displayed on a bar graph. Continuous, quantitative data usually requires a line graph. Graphs help scientists get a visual illustration of the data and allow them to analyze the data and recognize patterns.

How to Draw a Line Graph

1. On graph paper, draw a horizontal and vertical line, which intersect at a right angle. The horizontal line is called the x-axis, and the vertical line is called the y-axis. (See the figure.)

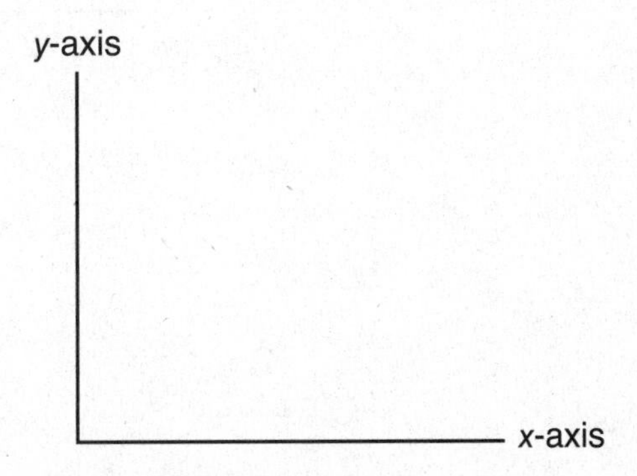

2. Examine the data that will be graphed and determine the independent (manipulated) and dependent (responding) variables. The independent variable will be recorded on the *x*-axis, and the dependent variable will be recorded on the *y*-axis. (See the figure.)

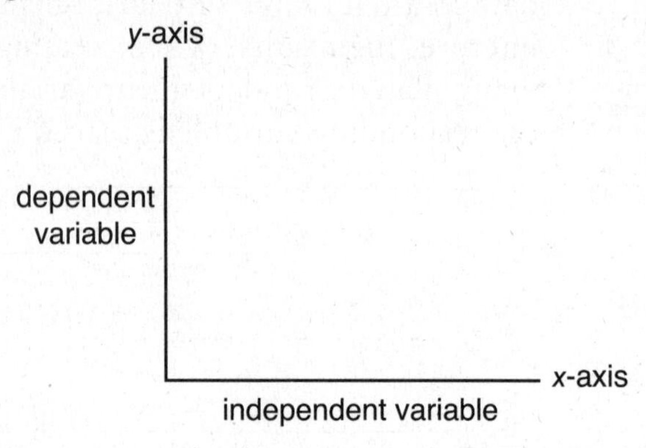

3. Determine how to number the increments along the *x*-axis and *y*-axis. Look at the range, highest and lowest numbers, in your data to help determine the increments that will be used for each axis. (See the figure.)

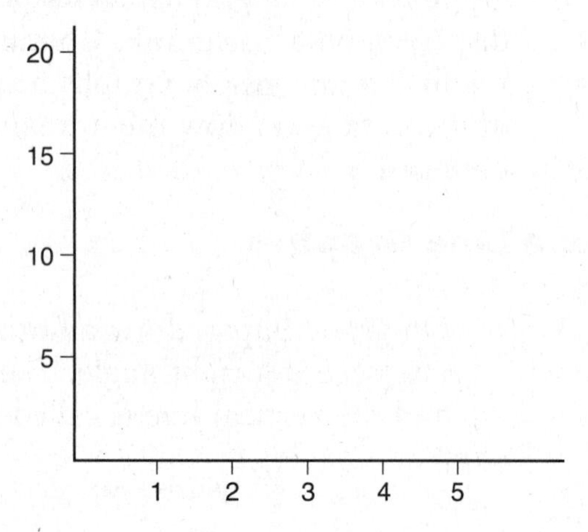

4. Plot your data. Then connect your data points on the line graph. (See the figure.)

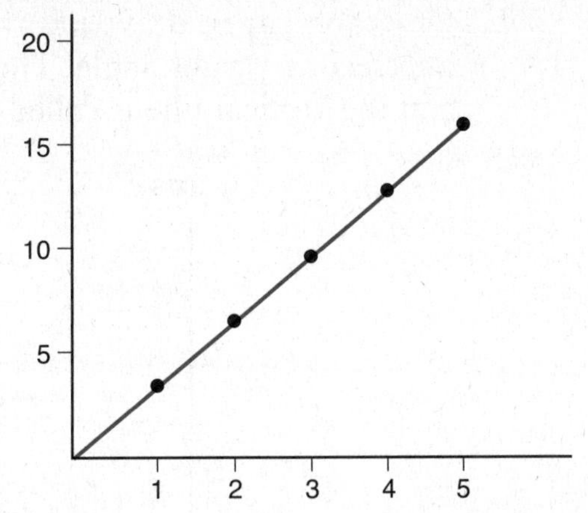

5. Be sure the graph is titled and each axis is labeled. Include units of measurement (grams, centimeters, etc.) (See the figure.)

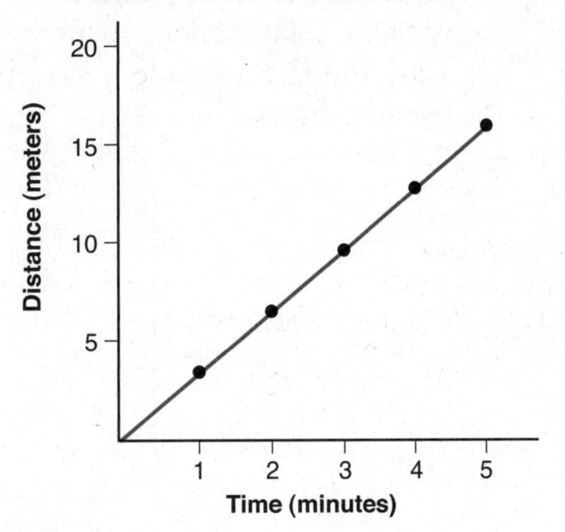

How to Draw a Bar Graph

1. On graph paper, draw a horizontal and vertical line, which intersect at a right angle. The horizontal line is called the *x*-axis, and the vertical line is called the *y*-axis. (See the figure.)

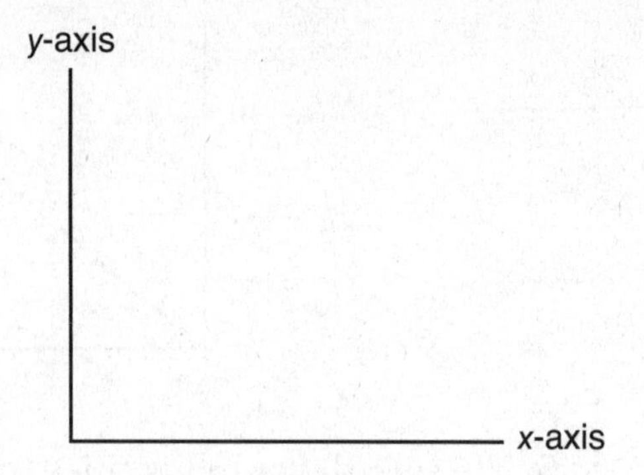

2. Examine the data that will be graphed and determine the independent (manipulated) and dependent (responding) variables. The independent variable will be recorded on the *x*-axis, and the dependent variable will be recorded on the *y*-axis. (See the figure.)

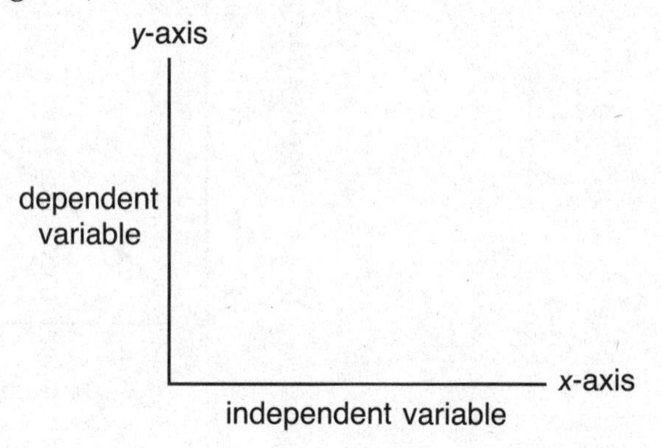

3. Determine how to number the increments along the y-axis. Look at the range, highest and lowest numbers, in your data to help determine the increments that will be used for the y-axis. (See the figure.)

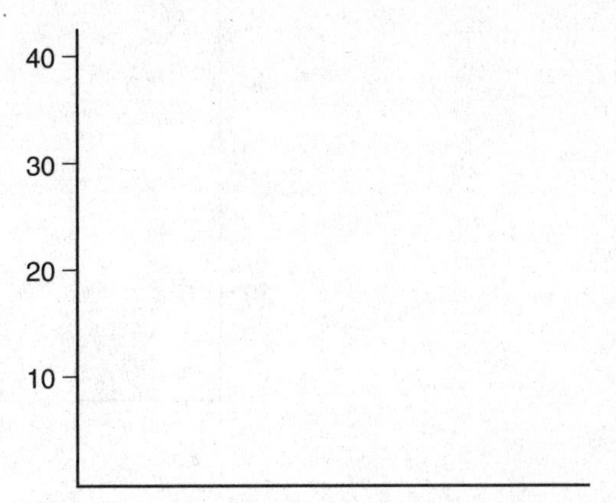

4. Write the labels of the categories for the dependent variable on the x-axis, and determine how wide each bar should be drawn. (See the figure.)

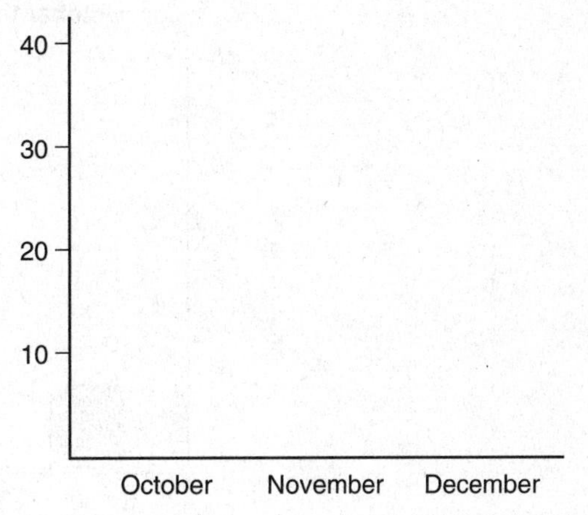

5. Plot your data so that the height of each bar corresponds with the appropriate measurement on the y-axis. (See the figure.)

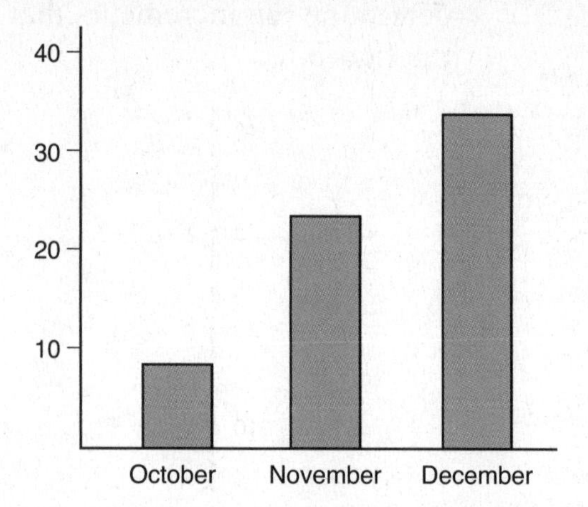

6. Be sure the graph is titled and each axis is labeled. Include units of measurement (grams, centimeters, etc.). (See the figure.)

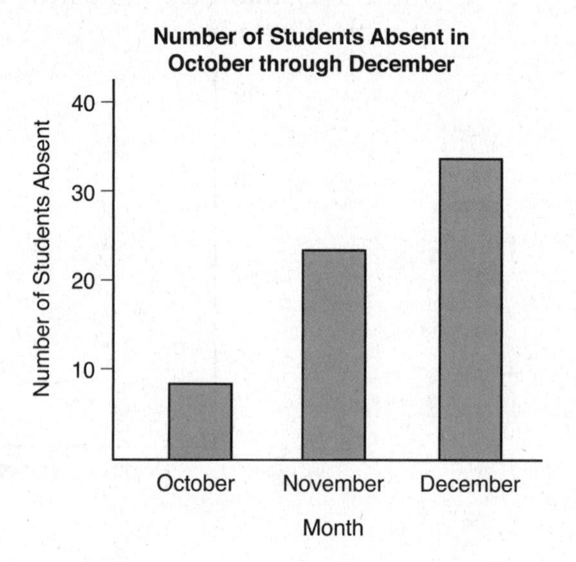

Example 5

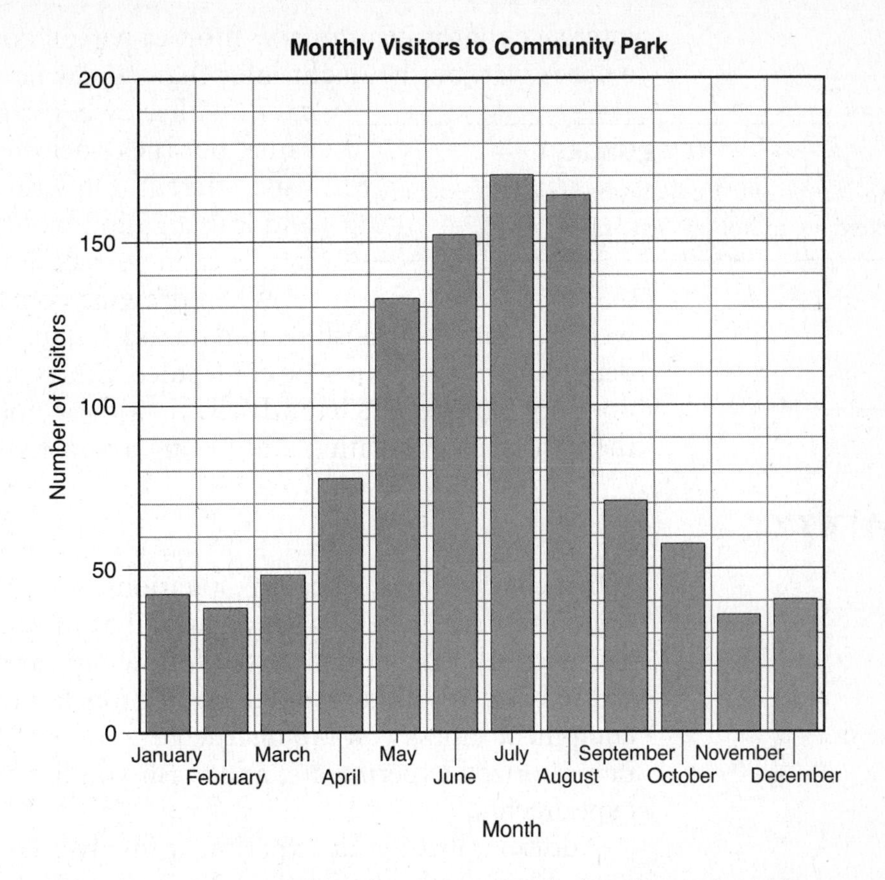

A worker at a community park recorded and graphed the number of monthly visitors to the park. Using the bar graph, which month does the park receive the most number of visitors?

A. February

B. July

C. August

D. November

See page 28 for the answer.

MAKING INFERENCES AND DRAWING CONCLUSIONS

Data tables and graphs help scientists make sense of data and observations, and they help determine whether there may be errors in the procedures of their investigation. Once the data and observations are organized, scientists can make inferences. An inference is a conclusion based on evidence, like data and observations. The same set of data can be interpreted in many ways. For example, if you observe that a car is wet, you might make the inference that it just rained. Another person could also observe that a car is wet, but he might infer that the owner had just washed the car. In either case, the observation is the same, but the conclusion each person made was different. Only through further study and investigation would you find out which inference in correct.

> **KEY CONCEPT**
>
> An inference is a conclusion based on evidence, like data and observations.

When drawing conclusions, evidence is weighed, and relationships in data are stated. Data is examined to determine if it supports or contradicts the hypothesis. If the data does not support the hypothesis, then these unexpected results are the first steps in creating and testing a new hypothesis.

SAFETY

When conducting science investigations, safety must always come first. There are general science rules that must always be followed in a science lab. These include the following: carefully follow all written and verbal directions, do not touch any materials or equipment unless you are instructed to do so, do not perform unauthorized experiments, and wash you hands before and after experiments.

Additionally, if your experiment involves heat, chemicals, living organisms, or glassware, then there are more specific safety rules you must follow. You should always wear safety goggles and tie back long hair. When using heat, you should also use heat-resistant gloves that protect your hands from hot glassware. Never handle broken or cracked glassware. Treat all living organisms and dissecting specimens in a humane manner. Also, you should know the location of safety equipment, like fire extinguishers and eye wash stations, in your classroom. Finally, never taste, touch, or smell any chemicals unless instructed to do so. Safety is always the top priority, and safety rules must always be followed.

Example 6

After conducting a laboratory frog dissection, what should students do for safety reasons?

A. Write a lab report.

B. Wash their hands with soap and water.

C. Clean the frog specimen with soap and water.

D. Read about the organ systems of amphibians.

See page 28 for the answer.

SCIENCE, SOCIETY, AND TECHNOLOGY

After reading this section you should be able to:

- Recognize that our science knowledge depends on the contributions of many scientists and from many different cultures.
- Describe the relationship between science and technology.

Our society is greatly impacted by the work of scientists from many cultures and from all around the world. Throughout history, scientific discoveries and technological breakthroughs have contributed to the advancement of mankind. The body of scientific knowledge continues to grow and develop because of the work of thousands of scientists.

THE ROLE OF SCIENCE IN OUR SOCIETY

Science is a human endeavor; it requires creativity, reason, and insight. Scientists work all around the globe to improve our lives. Doctors create new medicines and techniques to cure diseases. Biologists study the vast diversity of living things on our planet. Engineers create newer, better ways to construct buildings and build bridges. Scientists from different disciplines work together to solve problems. For example, doctors and engineers work together to design artificial, or prosthetic, body parts. Biologists work with chemists to understand the chemical reactions that occur within living organisms. By working together and sharing information, scientists contribute to a greater understanding of our world.

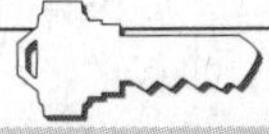

KEY CONCEPT

Science is a human endeavor; it requires creativity, reason, and insight. Scientists work all around the globe to improve our lives.

With scientific advancements come controversial issues. Citizens must decide their stance or position on a variety of debatable science topics like the genetic engineering of food, stem cells, alternative energy sources, and human cloning. It is the responsibility of all citizens to understand science so they are able to make informed decisions in their daily lives. Politics, religious values, and popular culture all steer science in the direction of the needs and wants of people. Scientists begin to study new areas and invent new technologies based on the perceived views and priorities of our society.

SCIENCE AND TECHNOLOGY

Science and technology work together to solve problems. Science attempts to explain the natural world, and technology attempts to provide tools and solutions. Technology solutions are usually temporary because advancements and improvements are always being made. New technologies not only provide benefits; there are usually trade-offs like costs and risks of new technology. Technology is essential to science because it provides instruments and tools for investigations. The quality of life and the interactions of people are influenced by technology.

GREAT TECHNOLOGICAL ACHIEVEMENTS IN SCIENCE

Science and technology change and evolve based on the needs and wants of society. Through science and technology, many inventions have revolutionized human life in the last one hundred years. For example, most of us take electricity for granted. As we turn switches on and off, electricity is delivered to us in order to power countless machines and appliance in our homes. Most of this electricity is delivered to us through wires from power plants that burn fossil fuels. Through energy transformations, the energy stored in fossil fuels (coal, oil, and natural gas) is changed to the electricity we use every day.

Automobiles transport us from place to place each day. Around 1900 there were only 10,000 or so cars in America. Now there are more than 60 million cars driven in the Unites States. Henry Ford revolutionized the process of assembling automobiles with the assembly line. This saved time and money by enabling many more automobiles to be built and distributed. Cars make it easier for us to travel long distances. The future of automobiles is exciting as scientists develop cars that can run on energy sources other than gasoline. Cars improve our ability to travel great distances, but airplanes allow us to travel around the world. As scientists improve their understanding of aerodynamics (the study of how air flows) and the study of flight, larger and faster airplanes are built.

Air conditioning and refrigeration change the way we eat and live. Without refrigerators, fresh foods would last two to three days at most. Refrigerators and refrigerated trucks allow for fresh foods to be available anywhere in the country. Homes and buildings used to be designed with natural cooling methods incorporated, like shading and cross-ventilation. Summers would be unbearable in many locations, especially the south. With the technology of air conditioning, the quality of life improved, and more people began to move to the warmer, southern states.

Because of the breakthrough of radio and television signals, there is mass communication and distribution of entertainment, news, and sports programming. Radio energy also allows police, fire departments, and pilots to communicate on assigned radio frequencies, or channels. Radio signals also carry information for cell phones and other wireless computer devices, changing the way we communicate and gather information.

Personal computers allow us to manage our lives and communicate via the Internet, and they have dramatically affected society. Computers complete enormous computing tasks for businesses, assist with robotics tasks for manufacturing, store and transfer data for financial companies, and more. In our homes, computers are used for gaming, shopping, learning, and communicating with others. The capability of computer technology is limitless as scientists develop faster and smaller computer components.

These are just a few of the greatest achievements in science and technology. They came to be because of the collective knowledge and work of many people throughout the world and throughout history. Cultural, political, and moral values and attitudes of our society influenced the development of and improvements made on these technologies.

Example 7

Henry Ford revolutionized both the industries of automobiles and manufacturing when he developed the assembly line. Which of the following is NOT a benefit of the assembly line?

A. It saves production time.

B. It saves money.

C. It increases the number of items manufactured.

D. It increases the knowledge of factory workers.

See page 28 for the answer.

LIVES AND CONTRIBUTIONS OF IMPORTANT SCIENTISTS

When examining the contributions of scientists throughout history, there are a few scientists who stand out because of the vast impact their work had on what we know about the world in which we live. It is not necessary to memorize the specific contributions of scientists, but it is important to understand and appreciate the context of their work and how it influenced society. Please note that throughout this manual, several scientists and their contributions will be discussed in the context of each science discipline, earth science, biology, and physical science. A few of the most influential are discussed in this section.

Isaac Newton (1642–1727) is considered by many to be the father of modern science. Not only did Isaac Newton discover the basic workings of the universe, but he also invented calculus, a complicated branch of mathematics. He helped define the laws of gravity and planetary motion. He also explained the laws of light and color using a prism.

Almost all of our understanding of living things on Earth depends on Charles Darwin's (1809–1882) theory of evolution by natural selection. Darwin's book, *On the Origin of Species*, took twenty-three years to publish. He worked on it during a five-year world voyage as he studied numerous fossils and living organisms. He concluded that species evolve as organisms with helpful traits survive, reproduce, and pass their traits on to future generations.

Louis Pasteur (1822–1895) contributed to the science of medicine when he developed the germ theory, which proposes that microorganisms are the cause of many diseases. He convinced many people and institutions of the value of hygiene and sanitation. And, he developed and used many vaccines, which saved hundreds of millions of lives.

These scientists along with thousands of others contribute to the body of scientific knowledge. Throughout history and all around the world, society has benefited from the work of the many scientists.

MATHEMATICAL APPLICATIONS

After reading this section you should be able to:

- Describe the tools and techniques of measurement in making observations.
- Represent and analyze data in data tables and graphs.

Just as science and technology are related to one another, so are science and mathematics. Science cannot be understood without using mathematics to express natural laws and represent information. Math is also used to record and compare measurements. Tables, graphs, and equations are all ways scientists use math to display information and show relationships between variables.

MEASUREMENT

When conducting investigations, scientists prefer to record quantitative observations in a systematic, or orderly, manner. Even though estimating may be worthwhile in some instances, scientists try to make accurate and precise measurements. Because scientists also want to communicate their results with others, a standard measuring system must be used. Otherwise, for example, one scientist may measure the length of specimen in feet, and another measures the length of a specimen in meters.

Throughout most of the world, the International System (SI) of Measurement is accepted as the standard of measurement. Four of the most common base units in SI are the second (for measuring time), the liter (for measuring volume of a liquid), the meter (for measuring length), and the kilogram (for measuring mass). A prefix may be used with the base unit name to indicate the size of the unit. Each prefix represents a multiple of ten. See the table for common prefixes.

SI Prefixes

Prefix	Symbol	Meaning	
kilo-	k	1000	thousand
hecto-	h	100	hundred
deka-	da	10	ten
deci-	d	0.1	tenth
centi-	c	0.01	hundredth
milli-	m	0.001	thousandth

The SI unit for length is a meter. A golf club is about one meter long. A meter can be divided into smaller units called centimeters and millimeters. A centimeter is one hundredth of a meter. So, there are one hundred centimeters in one meter. A millimeter is one thousandth of a meter. So, one meter is equal to one thousand millimeters.

Example 8

A student wants to add the same amount of water to her plant each day. To accurately measure the water, the volume of the water should be measured in

A. meters.

B. kilograms.

C. milliliters.

D. centimeters.

See page 28 for the answer.

Various tools are used to make measurements. No matter how careful a person is when measuring, there are always errors in measurements. Scientists select measuring instruments based on the degree of precision required. **Precision** refers to the ability of a measurement to be consistently reproduced. For example, you could test the precision of an electric scale by first placing an apple on the scale and recording the mass displayed. Then, remove the apple, and place it on the scale again. If the scale displays the same measurement for the apple's mass, then the scale is precise. Scientists desire their measurements to be accurate as well. Just because an instrument is precise does not mean that the measurement is accurate. **Accuracy** refers to how close the measurement is to its true value. Various factors can affect the accuracy of a measurement. For example, an instrument may be broken. One of the main factors that affects accuracy is human error. When making measurements, often the person may not know how to use and read the instrument properly. This can cause inaccurate measurements.

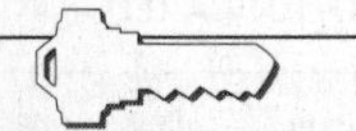

KEY CONCEPT

Precision refers to the ability of a measurement to be consistently reproduced. Accuracy refers to how close the measurement is to its true value.

One of the most common tools for measurement is the meter stick. A meter stick can be used to measure the length of an object using the meter base unit. You can also use the meter stick to determine the volume of rectangular solids. Volume is the amount of space occupied by an object. The volume of a rectangular solid can be determined by measuring the length multiplied by the width multiplied by the height of the object. The cubic meter (m^3) is the SI unit for the volume of a solid.

The volume of a liquid is measured using a graduated cylinder. The standard unit for the volume of a liquid is the liter. A **graduated cylinder** is a cylindrical container with markings along the side. The markings usually represent increments of milliliters. When reading a graduated cylinder it is important to view the surface of the liquid at eye level so that an accurate measurement is recorded. When some liquids, like water, are poured into a graduated cylinder, the surface of the liquid curves as the water clings to the sides of the container. This is called the **meniscus**. Be sure to read the marking beside the bottom of the concave meniscus to record an accurate reading. (See the figure.)

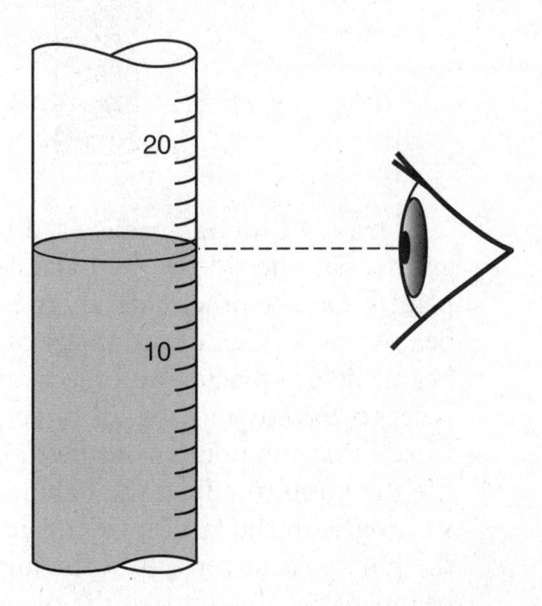

A graduated cylinder can also be used to determine the volume of irregularly shaped solids. For example, if you wanted to find the volume of a marble first find a graduated cylinder that is large enough for the marble to fit in. Fill the graduated cylinder with enough water to cover the marble. Before the marble is placed in the graduated cylinder, measure and record the amount of water within it. Let's say there are 40 milliliters of water in the graduated cylinder. Carefully drop the marble into the graduated cylinder. The water should rise because of the displacement by the marble. Measure and record the volume of the water level. Let's say that it reads 65 milliliters. Subtract the initial measurement from the final measurement to determine the volume of the marble. In this case, the marble has a volume of 25 milliliters, which is equivalent to 25 cm^3. This is called the **water displacement method**. (See the figure.)

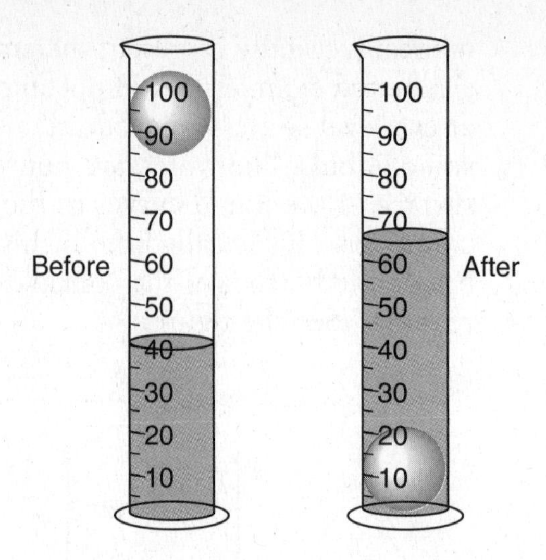

A **triple beam balance** is used to find the mass of an object in grams. On one side of the balance is a pan where the object is placed. On the other side, there is a set of three beams. Each of the beams has a rider, or an object of known mass, that slides across the beam. Before placing an object on the pan, be sure that the balance is set to zero by moving all three of the riders to the zero marking. Check that the pointer on the right is at the zero point. If it isn't, use the knob to adjust the balance. Place the object on the pan and, starting with the largest rider, slide the riders along the beams until the pointer is at zero. Add the masses indicated on each of the beams to get the mass of the object. (See the figure.)

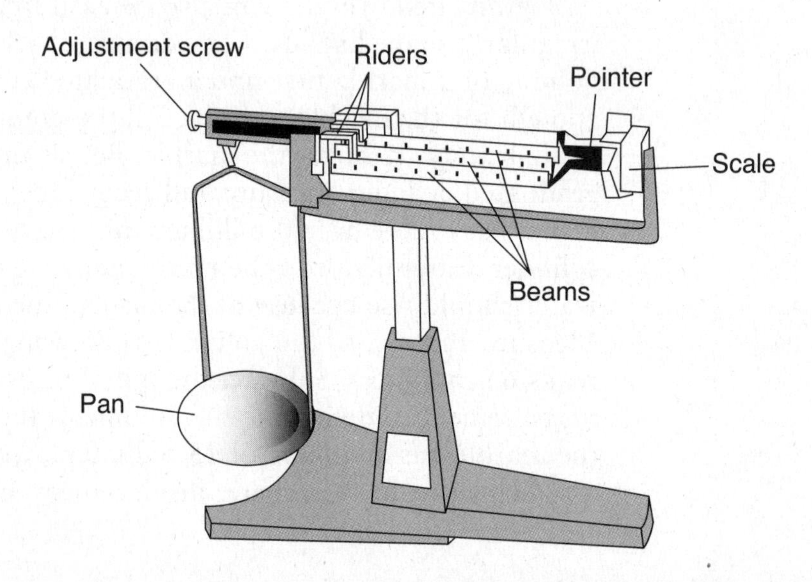

Other tools, thermometers and timers, are used to make additional measurements. As scientists make observations during investigations, these tools help them record data that can be compared and analyzed.

ANALYZING DATA IN TABLES AND GRAPHS

As discussed in "The Processes of Science," tables and graphs are used to organize and make visual representations of data. There are a variety of mathematical techniques that are used to analyze the data in tables and graphs.

There are a few ways to analyze the central tendency, or average, of a set of data. The mean is determined by dividing the sum of a set of data quantities by the number of quantities in the data set. The median is the middle value in a set of numbers. The mode is the most frequent value that appears in a set of numbers.

Let's look at an example. A student surveyed nine students in her class and recorded the number of siblings of each classmate. Her results were 0, 0, 1, 1, 2, 3, 3, 3, 4 siblings. To determine the mean of this data set, first add all of the quantities (0 + 0 + 1 + 1 + 2 + 3 + 3 + 3 + 4) and then divide by the number of students surveyed (9). The mean of the data set is 17 divided by 9, or 1.9. The median of this data set is the middle value of the set of numbers. In this case, there are nine numbers, so the fifth number in the data set is the median. The median of this data set is 2. Since the number 3 appears three times in the data set, it is the most frequent value and is called the mode.

When analyzing data in charts and graphs, predictions can be made by estimating values between known values or, in fact, beyond known information. For example, suppose you recorded the outside temperature every two hours as displayed in the table on page 24. If asked to estimate the temperature at 5:00 A.M., you could examine the pattern in the data and infer that the temperature was 22 degrees Celsius. This is called interpolation, or estimating a value between two values that are known. If asked to estimate the temperature at 10:00 A.M., you might examine the data you have and conclude that the temperature appears to rise by four degrees every two hours. Using this information, you can infer that the temperature at 10:00 A.M. will be 32 degrees Celsius. This is called extrapolation, or estimating a value beyond known values. Interpolation and extrapolation can be used with graphs as well as with data organized in tables.

Time and Temperature	
Time	Temperature
4:00 A.M.	20 degrees Celsius
6:00 A.M.	24 degrees Celsius
8:00 A.M.	28 degrees Celsius
10:00 A.M.	

Example 9

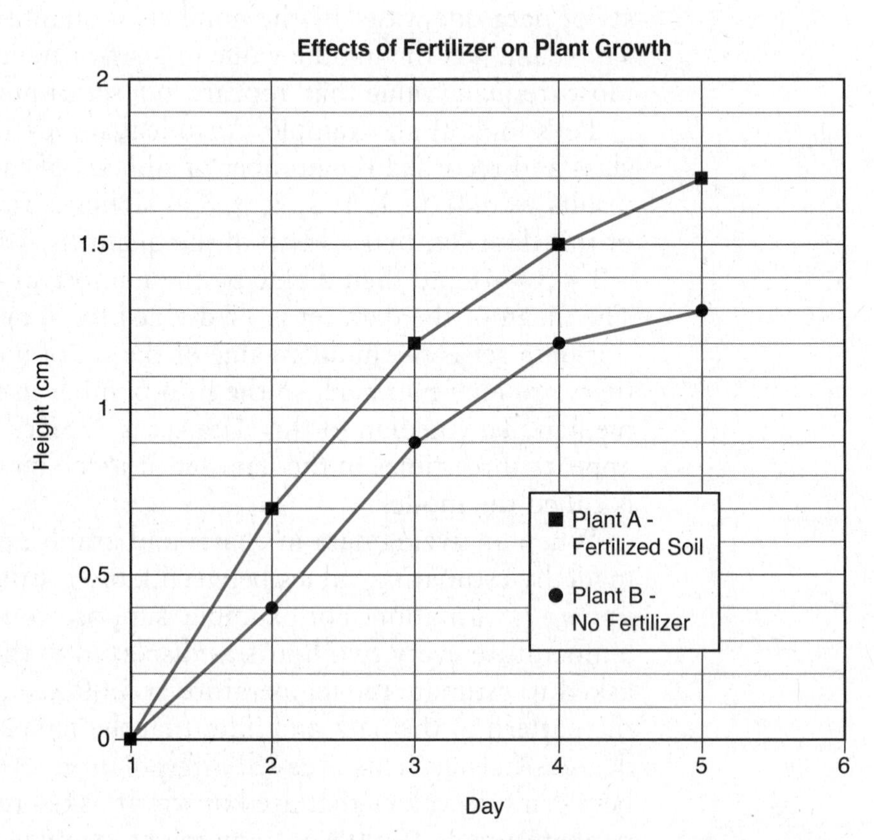

An experiment was conducted to determine the effect of fertilizer on the growth of plants. Based on the line graph, predict what the height of Plant A will be on day 6.

A. 1.0 centimeter

B. 1.5 centimeters

C. 1.8 centimeters

D. 2.5 centimeters

See page 28 for the answer.

Patterns and trends in data can allow scientists to analyze data and make predictions. When examining and analyzing graphs, patterns are easy to recognize. Take a look at the figure showing time versus distance. The line on the graph shows that as time increases, distance also increases; this is a direct relationship. Now take a look at the figure showing time versus temperature. This graph displays a line that shows that as time increases, temperature decreases; this is called an indirect relationship. By determining these patterns on graphs, the relationships between variables can be identified and communicated.

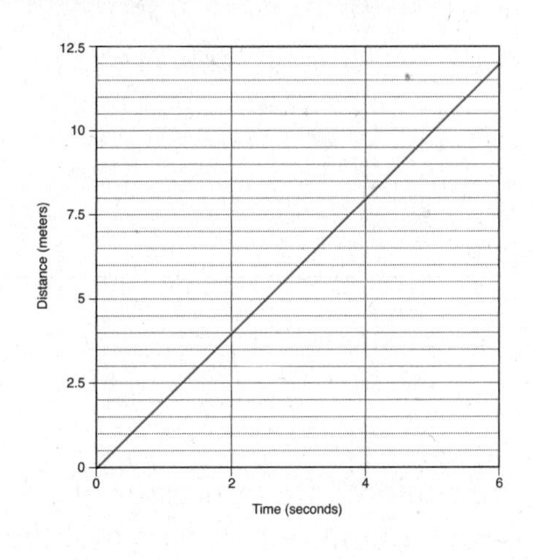

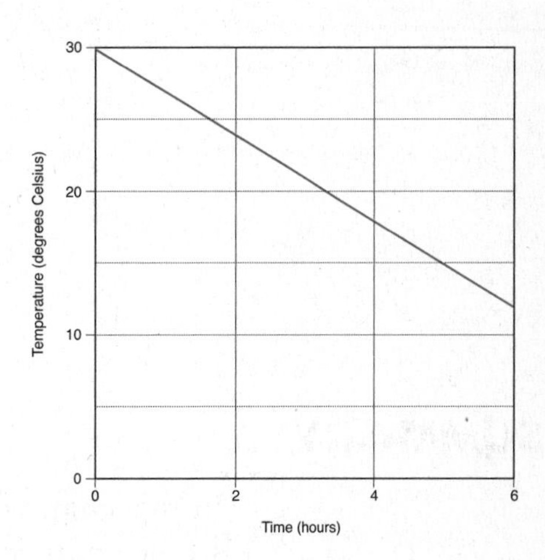

Example 10

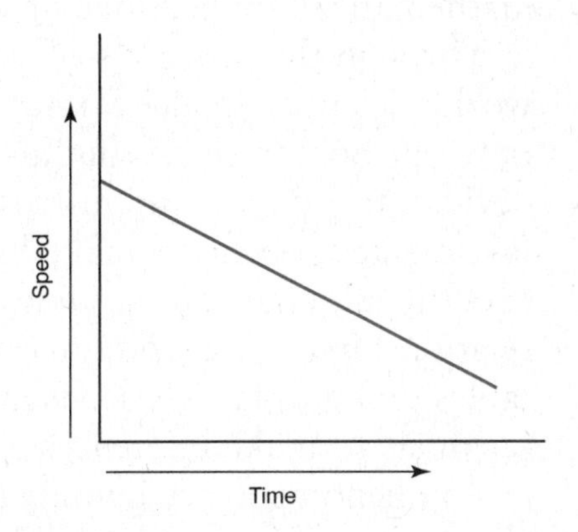

Which of the following statements best describes the relationship between speed and time in the line graph above?

A. There is no relationship between speed and time.

B. Speed remains constant over time.

C. Speed increases over time.

D. Speed decreases over time.

See page 28 for the answer.

Sometimes data points on a graph cannot be connected with a straight line. This is often because of errors in the data and observations. When this is the case, a **best-fit line** is drawn to represent the general pattern of the data. Some data points will fall above the line and some will fall beneath it. (See the figure.)

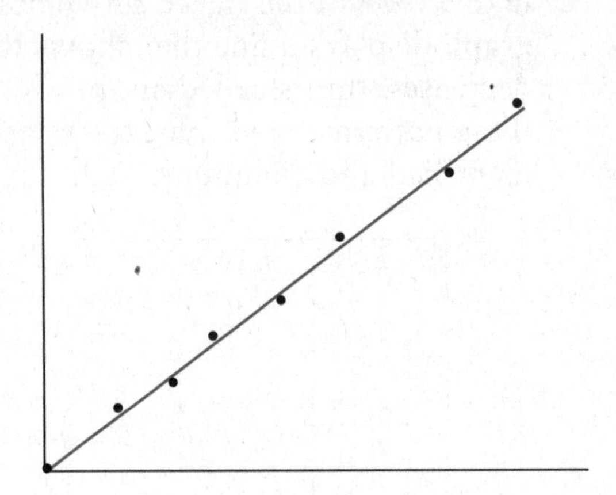

SUMMARY

In this chapter, we explored the processes of science, the role of science in technology and in our society, and the use of mathematical applications in science.

Through the processes of science inquiry, scientists study our world. Scientists make careful and detailed observations using not only our human senses but also many instruments, like measuring devices, telescopes, and microscopes. As scientific questions are investigated, scientists make hypotheses and design controlled experiments that test variables and look for cause-and-effect relationships. Data from investigations is organized and analyzed in tables and graphs. These visual representations of data enable scientists to make inferences and draw conclusions.

As phenomena are investigated in science laboratories and classrooms, safety is always the first priority. By following safety procedures and rules, and by knowing how to use safety equipment, meaningful science investigations can be conducted.

Technological innovations and new scientific knowledge can be credited to countless scientists who lived throughout history and who came from a variety of cultures. The political, moral, and social views of citizens drive science and technology to solve problems. Technology attempts to create solutions, but there are always risks and trade-offs associated with new technology. Technological breakthroughs, like refrigeration and new transportation options, dramatically changed the quality of human life and our ability to live in and explore our world.

Mathematics is also a crucial part of answering questions about our world. Math can be used to make observations and measurements, organize data, and create arguments and explanations. Scientists use a standard of measurement called the International System of Measurement. This ensures that measurements can be compared. There are a variety of instruments that scientists use to make accurate and precise measurements.

Data is organized and displayed in data tables and graphs. The data can be interpreted and analyzed by determining central tendencies and through extrapolation and interpolation. This analysis assists scientists in making predictions and drawing conclusions.

KEY TERMS FROM THIS CHAPTER

phenomena
inquiry
observations
data
instrument
qualitative
quantitative
variable
independent variable
manipulated variable
controlled variable
constants
dependent variable
responding variable
hypothesis
controlled experiment
control
repeated trials
data table
graph
line graph
bar graph
inference
International System (SI) of Measurement
second
liter
meter
kilogram
precision
accuracy
graduated cylinder
meniscus
water displacement method
triple beam balance
central tendency
mean
median
mode
interpolation
extrapolation
direct relationship
indirect relationship
best-fit line

ANSWERS TO EXAMPLES

1. **D** The other responses are questions about personal preferences and cannot be answered through a scientific investigation.
2. **D** The independent variable is changed purposefully, and in this investigation it is the amount of water. Response A is the dependent variable, and responses B and C are constants.
3. **C** Response C is a hypothesis written in the "if . . . then . . . " format. Responses A and D are observations, and response B is a question.
4. **A** To check the results of the experiment, the experiment should be repeated.
5. **B** July had the most visitors (170) during a month.
6. **B** Always wash your hands after an experiment that uses chemicals or living specimens.
7. **D** Assembly line workers actually know less about the entirety of the manufactured item. Their jobs are usually very specialized.
8. **C** The standard for measuring the volume of a liquid is the liter. A milliliter is one thousandth of a liter.
9. **C** You can extrapolate that Plant A will be 1.8 centimeters tall on day six.
10. **D** The graph shows an indirect relationship between speed and time whereby speed decreases as time passes.

PRACTICE QUIZ

For each of the questions or incomplete statements below, choose the best of the answer choices given. Turn to page 235 for the answers.

1. Which of the following questions cannot be answered through a scientific investigation?

   A. How does air temperature affect air pressure?

   B. How does salt affect the melting point of ice?

   C. How does music inspire people?

   D. What is the relationship between elevation and air pressure?

2.

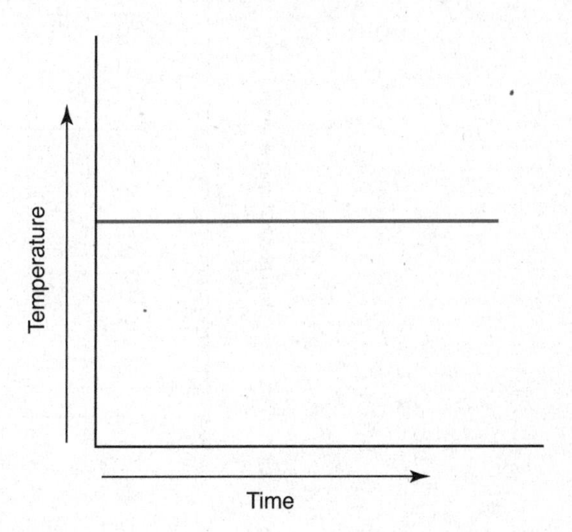

   Which of the following statements best describes the relationship between temperature and time in the line graph above?

   A. There is no relationship between temperature and time.

   B. Temperature remains constant over time.

   C. Temperature increases over time.

   D. Temperature decreases over time.

3. A student wanted to test how temperature affects the sprouting of alfalfa seeds. She set up five dishes and placed ten alfalfa seeds on a moist paper towel in each dish. Each of the five dishes was kept at a different temperature—10°C, 15°C, 20°C, 25°C, and 30°C. Which of these is the independent, or manipulated, variable in this experiment?

   A. the number of seeds

   B. the number of dishes

   C. the temperature of each dish

   D. the moist paper towel

4. 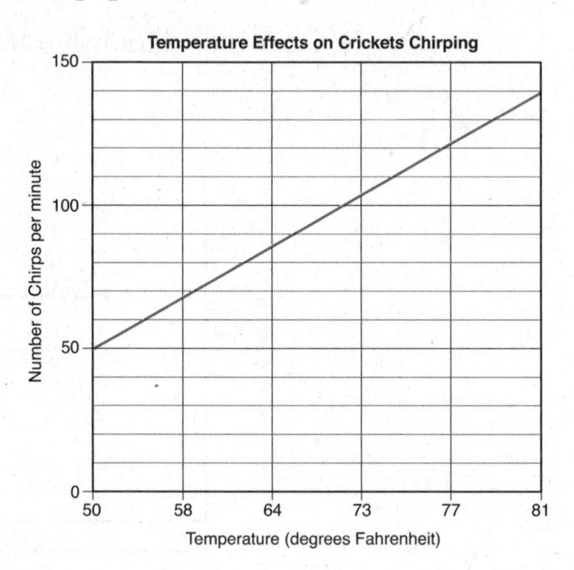

   A science class investigated whether there was a relationship between the temperature and frequency of chirping in crickets. What analysis can be made of their data represented in the line graph?

   A. There is no relationship between temperature and the frequency of a cricket's chirps.

   B. As the temperature increases, the chirping cricket becomes louder.

   C. As the temperature increases, the number of chirps per minute increases.

   D. As the temperature increases, the number of chirps per minute decreases.

5. A student wondered if eating breakfast improves your grades in school. Which of the following best represents a possible hypothesis from this student?

    A. Eating breakfast has no effect on attendance at school.

    B. If students eat breakfast, then grades will improve in school.

    C. If grades improve in school, then students will enjoy breakfast.

    D. How does eating breakfast affect a student's grades?

6. What instruments can be used to measure the volume of a small rock?

    A. a spring scale and a ruler

    B. a barometer and a thermometer

    C. a graduated cylinder and water

    D. a triple beam balance and water

7. A team of students examined an object. Which of the following is NOT an observation about the object?

    A. It is smooth and shiny.

    B. It has a mass of 552 grams.

    C. It is 21 centimeters long.

    D. I think it is metal and used in a machine.

8.

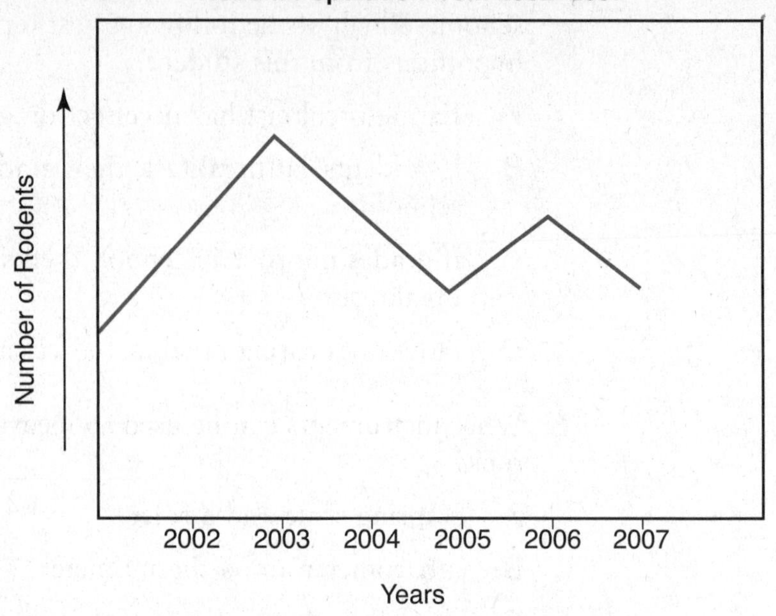

Barn owls primarily feed on rodents. Using the line graph, in which year was there the least amount of competition among barn owls for rodents?

A. 2002

B. 2003

C. 2005

D. 2006

9. Two beakers were placed on a windowsill. One beaker contained 100 milliliters of water, and the other beaker contained 100 milliliters of rubbing alcohol. The next day both cups had less liquid in them, but there was less rubbing alcohol than water. What can be concluded from this experiment?

A. Only some liquids evaporate.

B. Liquids can only evaporate in the presence of sunlight.

C. Some liquids evaporate faster than others.

D. Water evaporates faster than rubbing alcohol.

10. Daily Measurement of the Height of a Plant

Day	Height
Monday	0 cm
Tuesday	0.2 cm
Wednesday	0.4 cm
Thursday	0.6 cm
Friday	0.8 cm
Saturday	
Sunday	
Monday	1.4 cm

A student measured the growth of a plant each day in school. Because he did not go to school on Saturday and Sunday no measurement of the plant was made. Which of the following measurements would be a reasonable estimation of the height of the plant on Sunday?

A. 0.8 cm

B. 0.9 cm

C. 1.2 cm

D. 1.4 cm

11. Students worked in teams in a science class to complete the same experiment. After conducting the investigation and collecting data, one team gathered all of the data from the other teams. After comparing their data with the other teams' data, what would indicate that their data is valid?

A. Many of the other teams recorded similar data.

B. Their team finished before the other teams.

C. Another class completed the same experiment.

D. Their team followed all of the rules and procedures.

12. Which of the following statements is a hypothesis?

    A. Why do your ears pop in an airplane?

    B. Deer are herbivores.

    C. If the temperature of the water increases, then more sugar can dissolve in it.

    D. How do earthworms react in moist environments?

13. For a project on volcanoes, a student builds a 1:1,000 scale model of Mt. St. Helens. Since Mt. St. Helens is 1,550 meters tall, what is the height of the model?

    A. 0.155 meters

    B. 1.55 meters

    C. 15.5 meters

    D. 155 meters

14.

TIME AND TEMPERATURE

Time	Temperature
6:00 A.M.	45 °F
8:00 A.M.	49 °F
10:00 A.M.	52 °F
12:00 P.M.	58 °F
2:00 P.M.	62 °F
4:00 P.M.	57 °F
6:00 P.M.	52 °F

The data table shows the outdoor temperature at different times during one day. When was the highest temperature recorded?

A. 6:00 a.m.

B. 12:00 P.M.

C. 2:00 P.M.

D. 6:00 P.M.

15. Which instrument would most accurately and precisely measure the mass of an apple?

A. triple beam balance

B. spring scale

C. thermometer

D. graduated cylinder

16. Which of the following technological breakthroughs had the greatest impact on food consumption and distribution?

A. airplanes

B. computers

C. television and radio

D. air conditioning and refrigeration

17.

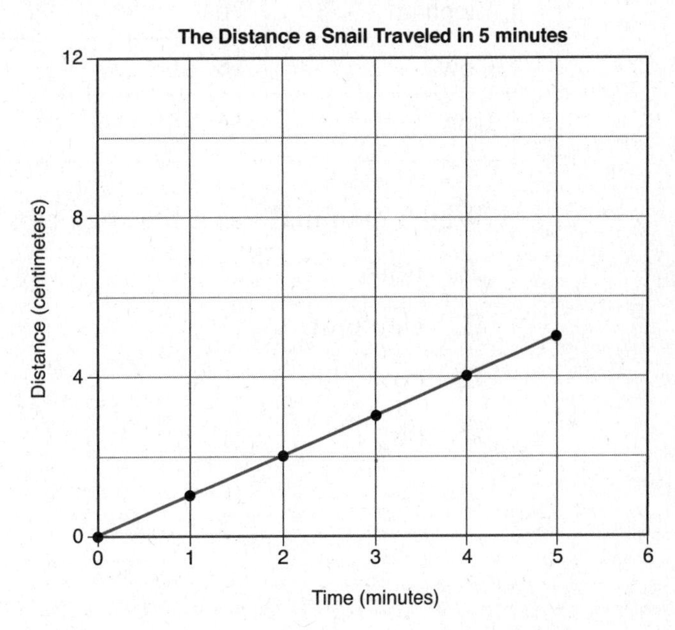

A student observed a snail and measured how far it traveled in five minutes. If the student continued to observe the snail, how far would it probably have traveled in six minutes?

A. 10 cm

B. 8 cm

C. 6 cm

D. 4 cm

18. Which of the following pieces of equipment would be useful for a scientist studying birds in their natural habitats?

A. binoculars

B. radio telescopes

C. microscopes

D. magnifying glasses

19. Lifespan and Gestational Period of Common Mammals

Mammal	Average Lifespan (years)	Gestational Period (days)
Human	77	266
Horse	27	330
Elephant	70	645
Cow	18	284
Dog	16	61

Which mammal has a life span closest to that of a human?

A. horse

B. elephant

C. cow

D. dog

20.

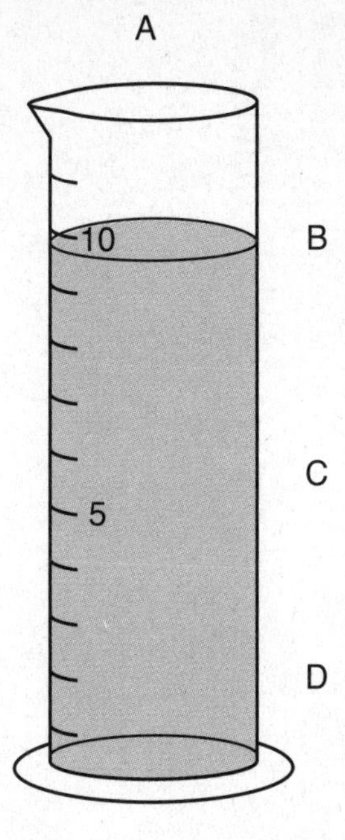

A student wants to read a graduated cylinder to determine the volume of a liquid. From which position should the student read the graduated cylinder to make the most accurate measurement?

A. position A

B. position B

C. position C

D. position D

Chapter 2

CHARACTERISTICS OF LIFE

Why does your hair turn gray when you age? Why are more people right-handed than left-handed? How can you clone an animal? Why is a spider's silk so strong? Biology is the study of life, and biologists work to answer these questions. In this chapter, we will review the diversity, complexity, and interdependence of life on Earth. We will also explore how organisms evolve, reproduce, and adapt to their environments. This chapter is divided into three sections:

- Matter, Energy, and Organization in Living Systems
  - Cells and Cell Parts
  - How to Use a Compound Microscope to Observe Cells
  - Cell Processes: Photosynthesis and Respiration
  - Levels of Organization
  - Organ Systems
- Diversity and Biological Evolution
  - Classification
  - The Six Kingdoms of Living Things
  - The Classification of Animals
  - Classes of Vertebrates
  - Naming Organisms
  - Acquired vs. Inherited Characteristics
  - Natural Selection, Evolution, and Extinction
- Reproduction and Heredity
  - Life Cycles
  - Growth and Reproduction
  - Genes, Chromosomes, and DNA
  - Dominant and Recessive Traits

NOTE TO STUDENTS

Thoroughly read each section. As you read, pay close attention to the key concepts and try to answer the example questions found in each section. Answers to examples in this chapter are on page 65. At the end of the Chapter 2, assess your knowledge of the life sciences with the practice quiz. Then, check your answers using the answer key provided on page 236.

MATTER, ENERGY, AND ORGANIZATION IN LIVING SYSTEMS

After reading this section you should be able to:

- Identify and describe the structure and function of cells and cell parts.
- Explain how the products of respiration and photosynthesis are recycled.
- Recognize that complex multicellular organisms, including humans, are composed of and defined by interactions of the following: cells, tissues, organs, systems.
- Explain how systems of the human body are interrelated and how they regulate the body's internal environment.

We live on Earth, a planet where life flourishes. Living things, or **organisms**, are all around us. Birds fly in the sky above. Dust mites live in our beds. There are even microscopic organisms living in our eyelashes. Life has been discovered in the most extreme conditions on Earth. Bacteria live in the boiling water of geysers and hot springs. Miles below the ocean surface and without any sunlight, fish, crabs, and giant worms live strange and fascinating lives. Life on Earth is quite diverse, but what do all living things have in common? What is life and what does it mean to be alive? Scientists have difficulty defining "life," but they can agree on five characteristics that all living things exhibit. All organisms:

1. are made of one or more cells
2. need nutrients to obtain energy
3. grow and develop
4. have the ability to reproduce
5. respond and adapt to changes in their environment

An organism must show evidence of all of these characteristics to be considered alive. Within this section, we will explore the first two characteristics listed above. The remaining characteristics will be discussed in the last two sections. Let us first examine the basic component of organisms, cells.

CELLS AND CELL PARTS

Cells are the tiny building blocks of living things. This explains why they are called the basic units of structure in organisms. Some organisms, like bacteria and amoebas, are made of just one cell. Other organisms, like animals and plants, are made of many cells. Cells provide a place for chemical reactions that maintain life to occur. Within cells, food is turned into energy. Many life functions, like reproduction, respiration, and even bone repair, take place at the cellular level.

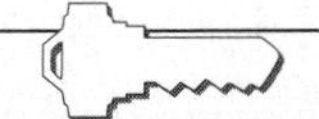

KEY CONCEPT

Cells are the basic units of structure and function in organisms.

There are many different types of cells, like plant and animal cells. Each type of cell varies in its structure and function. This means that the cells have different cell parts and different jobs. Some basic cell parts are the cell membrane, nucleus, mitochondria, cytoplasm, and vacuoles. The **cell membrane** is the thin, skin-like structure that surrounds the cell. It acts as a barrier and allows some materials, like water, to enter and exit the cell. The **nucleus** is like the brain of the cell because it controls all of the cell's activities and contains the genetic information for the organism. **Mitochondria** are sometimes called the "powerhouse of the cell" because, through the process of respiration, they provide the cell with energy. The nucleus and mitochondria are found in the **cytoplasm**, the jelly-like material that fills the cell. Cytoplasm is made mostly of water and is located between the cell membrane and the nucleus. It holds all of the cell parts in place and helps to maintain the shape of the cell. **Vacuoles** are storage bubbles. They act like a suitcase, storing food, water, and even waste for the cell. Vacuoles are much larger in plant cells than in animal cells. (See the figures.)

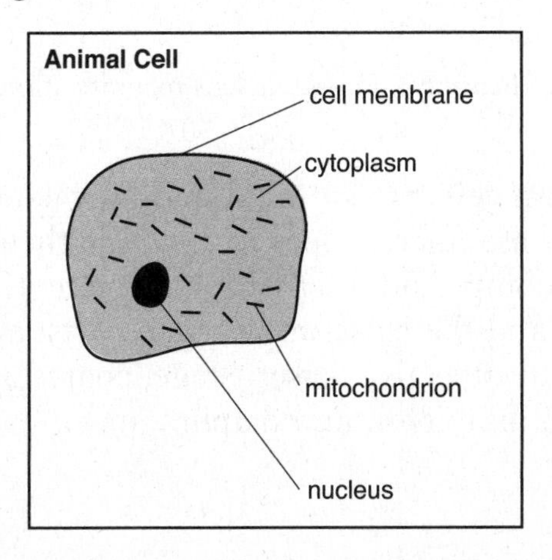

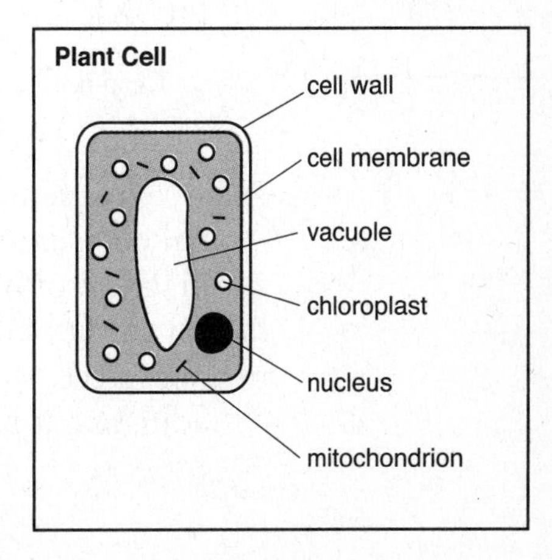

Plant cells have two important structures that animal cells do not have: the cell wall and chloroplasts. The cell wall provides a framework or rigid, tough structure outside the cell membrane. The cell wall is what makes wood and plant stems sturdy. Chloroplasts are the food producers of the cell and contain a pigment called chlorophyll, which provides plants with their green color. Most importantly, they play a role in the process of photosynthesis.

Example 1

The function of the cell nucleus is to

A. store food, water, and waste.

B. provide energy for the cell.

C. control the activities of the cell.

D. help the cell expel waste.

See page 65 for the answer.

HOW TO USE A COMPOUND MICROSCOPE TO OBSERVE CELLS

In 1665, a physicist named Robert Hooke discovered cells using a simple microscope. The basic microscope that Hooke used evolved into the modern compound microscope. A compound microscope uses more than one lens to magnify an object. There is a lens in the eyepiece and usually three lenses on a revolving nosepiece.

To determine the total magnification power of the microscope, multiply the eyepiece lens magnification by the objective lens magnification. For example, if the eyepiece has a magnification of ten times (10X) and the objective lens has a power of forty times (40X), then the total magnification equals four hundred times (400X).

Eyepiece magnification	×	Objective lens magnification	=	Total magnification
10X	×	40X	=	400X

Because most cells are so small that they cannot be seen with the naked eye, we use microscopes to examine them. Refer to the diagram of a compound microscope on page 43 to review the steps for using one and the basic microscope parts: eyepiece, revolving nosepiece, three objective lenses, stage, coarse adjustment knob, fine adjustment knob, mirror, and diaphragm.

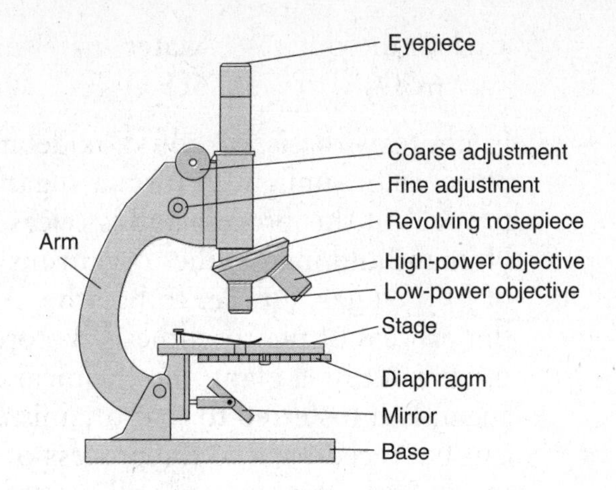

Steps for Using a Compound Microscope

1. Place the microscope on a solid, sturdy surface.
2. Turn the revolving nosepiece to the low-power objective lens (smallest lens).
3. Hold the specimen slide on the stage with the stage clips.
4. While looking at the microscope from the side, use the coarse adjustment knob to lower the lens as close as it will go to the slide without touching it.
5. Look into the eyepiece and adjust the mirror and diaphragm to allow the most amount of light in.
6. Slowly turn the coarse adjustment knob so that the lens moves away from the stage until the specimen comes into focus. Use the fine adjustment knob for fine focusing.
7. Move the slide until the specimen is in the center of the field of view.
8. Change to the next objective lens on the revolving nosepiece if desired for greater magnification.

CELL PROCESSES: PHOTOSYNTHESIS AND RESPIRATION

With the technology of microscopes, scientists observed that cells aren't just the building blocks of organisms; many processes occur within cells. As mentioned earlier, all organisms need nutrients to gain energy. Two processes involved with obtaining energy are photosynthesis and respiration. Both processes occur within cells. Through the process of **photosynthesis**, chloroplasts convert light energy from the sun into chemical energy stored as sugar. We can represent the process of photosynthesis by using the following chemical equation:

carbon dioxide	+	water	+	sunlight	→	sugar	+	oxygen
$6CO_2$	+	$6H_2O$	+	sunlight	→	$C_6H_{12}O_6$	+	$6O_2$

In photosynthesis, carbon dioxide and water combine in the presence of sunlight to form a sugar called glucose. Oxygen is a by-product of this process and is released as gas into the atmosphere. Humans and many other organisms are dependent upon this release of oxygen gas in order to breathe. Also, the sugar that is created in the process of photosynthesis is stored in the plant. When an organism eats a plant, the chemical energy, stored in the form of sugar, is transferred to that organism. All energy in an ecosystem can be traced back to the process of photosynthesis and ultimately to the Sun. This concept will be very important when we review the flow of energy in ecosystems in Chapter 3.

Respiration can be defined as the process by which an organism takes in oxygen and releases carbon dioxide. Let's examine respiration at the cellular level. Respiration is essentially the opposite of photosynthesis. In cells, mitochondria use oxygen and sugar to produce energy, carbon dioxide, and water; this is called cellular respiration. The chemical equation for respiration is

sugar	+	oxygen	→	carbon dioxide	+	water	+	energy
$C_6H_{12}O_6$	+	$6O_2$	→	$6CO_2$	+	$6H_2O$	+	energy

Photosynthesis occurs mostly in green plants, but respiration occurs in all organisms. All organisms require a continuous supply of energy to perform their life functions. During respiration, the chemical energy stored in glucose sugar is released. The carbon dioxide released during respiration can be "recycled" by plants during photosynthesis. Also, cells can use the sugar and oxygen produced through photosynthesis for respiration.

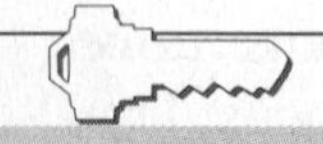

KEY CONCEPT

Carbon dioxide produced during respiration can become a raw material for photosynthesis. Likewise, the oxygen produced by photosynthesis becomes available for respiration.

Example 2

What is needed for the process of photosynthesis?

A. darkness, soil

B. darkness, oxygen, and glucose

C. light, carbon dioxide, and water

D. light, amino acids, and water

See page 65 for the answer.

LEVELS OF ORGANIZATION

Living things are well-organized, and they are all built with cells. Some organisms, like bacteria, are single-celled, or **unicellular**. **Multicellular** organisms, like humans, are made of many cells. In fact, the human body is made of over ten trillion cells! In multicellular organisms, cells vary in size, shape, structure, and function. Cells have different jobs. For example, in the human body, red blood cells carry oxygen around the body, and nerve cells send messages to the brain. Similar cells join together to form **tissues** that perform a special function. Different tissues join together to form **organs**, like the liver and heart. These organs also perform specialized functions. For example, the stomach produces acid that helps to digest food; the lungs transfer the oxygen from the air we breathe to our bloodstream. Organs work together in **organ systems** to carry out major bodily functions, like digestion and circulation.

Example 3

Which of the following represents the body structures from least to most complex?

A. cell, tissue, organ, organ system

B. tissue, organ, cell, organ system

C. organ system, organ, tissue, cell

D. cell, organ system, tissue, organ

See page 65 for the answer.

ORGAN SYSTEMS

Complex organisms, like humans, have a wide variety of organs that function together to maintain life. These organisms depend upon the performance of organ systems in order to survive. Therefore, if one organ is not working properly, the whole system is disrupted. You may experience a "stomachache," when in fact the problem is not in your stomach, but in your intestines. This intestinal problem may impact your entire digestive system.

KEY CONCEPT

Similar cells join together to form tissue. Different tissues combine to form organs. Organs are grouped into systems. Organ systems work together to maintain life.

Example 4

Which of the following best demonstrates the relationship between the respiratory and the circulatory systems?

A. The lungs provide energy to the blood.

B. The lungs provide water to the blood.

C. The lungs provide carbon dioxide to the blood.

D. The lungs provide oxygen to the blood.

See page 65 for the answer.

Humans have separate organ systems for digestion, respiration, reproduction, circulation, excretion, movement, and protection from disease. As we already know, these systems interact with one another. For example, the respiratory and circulatory systems cooperate to deliver oxygen and to remove carbon dioxide from the body. The skeletal and nervous systems work together as the skull protects the brain from injury. Refer to the table to review the organs and functions of the major organ systems in the human body.

MAJOR ORGAN SYSTEMS OF THE HUMAN BODY

Organ System	Major Organs	Function of Organ System
Digestive	Mouth, esophagus, stomach, intestines, liver, pancreas	Breaks down and absorbs nutrients from food
Circulatory	Heart, blood, blood vessels	Moves blood and nutrients throughout body to cells
Respiratory	Nose, trachea, lungs, bronchioles	Absorbs oxygen from air and expels carbon dioxide
Nervous	Brain, spinal cord, nerves	Transports electrical nerve signals to and from the brain and other parts of the body
Skeletal	Bones, cartilage, tendons, ligaments	Provides support for the body and protects internal organs
Muscular	Muscles	Provides movement
Reproductive	Female—ovaries, uterus, vagina Male—testes, penis	Manufactures cells that allow and support reproduction

Example 5

The esophagus, stomach, and intestines function together as components of which of the following?

A. system

B. web

C. tissue

D. organ

See page 65 for the answer.

Example 6

In humans, which two organ systems work to break down food and deliver nutrients to body cells?

A. respiratory and nervous

B. digestive and circulatory

C. nervous and digestive

D. respiratory and circulatory

See page 65 for the answer.

DIVERSITY AND BIOLOGICAL EVOLUTION

After reading this section you should be able to:

- Describe and give examples of the major categories of organisms and of the characteristics shared by these organisms.
- Compare and contrast between acquired and inherited characteristics.
- Compare and contrast organisms using their internal and external characteristics.
- Discuss how changing environmental conditions can result in evolution or extinction of a species.

CLASSIFICATION

There are millions of different organisms living on earth and millions more that previously existed but are now extinct. To make sense of the vast number and diversity of the organisms on Earth, scientists classify and name organisms. Biologists classify organisms based on similarities in their external and internal structures.

KEY CONCEPT

Organisms are classified based on the similarity of their physical structures.

Think about how you would classify the following animals into two groups: cow, starfish, wasp, salmon, snake, and lobster. Perhaps you would place the cow, salmon, and lobster in one group because they are all animals that you like to eat for dinner! But where does that leave our wasp, starfish, and snake? (Safe from the oven, I guess.) Even though this is one way to classify these animals, scientists group organisms according to their physical structures. One way to classify these animals based on their physical structures would be by whether or not they have a backbone. Cows, salmon, and snakes have a backbone and are called vertebrates. Starfish, wasps, and lobsters do not have a backbone and are called invertebrates.

Example 7

Biologically, organisms are classified based on their

A. acquired similarities

B. structural similarities

C. habitats

D. biochemical similarities

See page 65 for the answer.

Over time, scientists from around the world developed a biological classification system. This classification system is still growing and changing as scientists discover new organisms and reevaluate how organisms are related to one another. The complete classification system, from broadest to most specific grouping, includes kingdom, phylum, class, order, family, genus, and species.

THE SIX KINGDOMS OF LIVING THINGS

KEY CONCEPT

The biological classification system moves from broad groupings in kingdoms to more specific groupings of genus and species.

The largest and most general groupings are called kingdoms. The six kingdoms of living things are archaebacteria, eubacteria, protists, fungi, plants, and animals. When an organism is discovered, it is first classified into one of these six kingdoms based on the following criteria:

1. How many cells it has
2. What type of cells it has
3. How the organism obtains food

As already reviewed, an organism is either unicellular or multicellular. Next, an organism is either a prokaryote or a eukaryote. If the cell or cells of the organism do not contain a nucleus, then the organism is a **prokaryote**. If the cell or cells contain a nucleus, then it is a **eukaryote**. Lastly, if an organism produces its own food from the sunlight and other nutrients, then it is called an **autotroph**, and if an organism cannot produce its own food and relies on other organisms for nutrients, then it is called a **heterotroph**. Humans are heterotrophs because we ingest and consume other organisms for nutrients. Fungi, like mushrooms and molds, are also heterotrophs, but they absorb nutrients rather than ingest them. Review the characteristics of organisms in the six kingdoms in the table.

THE SIX KINGDOMS OF LIVING THINGS

Kingdom	Examples	Number of Cells	Cell Type	Food Source
Archaebacteria	Ancient bacteria	Unicellular	Prokaryotes	Autotrophs
Eubacteria	Bacteria like *E. coli* and salmonella	Unicellular	Prokaryotes	Some autotrophs, some heterotrophs
Protists	Euglena, amoebas, and paramecia	Most are unicellular	Eukaryotes	Some autotrophs, some heterotrophs
Fungi	Molds, mildews, mushrooms	Most are multicellular	Eukaryotes	Heterotrophs by absorption
Plants	Mosses, ferns, trees, flowering plants	Multicellular	Eukaryotes	Autotrophs
Animals	Jellyfish, worms, insects, mammals, reptiles	Multicellular	Eukaryotes	Heterotrophs by ingestion

Example 8

Which of the following levels of the biological classification system contain living things with the most characteristics in common?

A. species

B. class

C. family

D. kingdom

See page 65 for the answer.

THE CLASSIFICATION OF ANIMALS

After an organism is classified as an animal, it is then placed into a smaller, more specific grouping called a **phylum**. The animals within each phylum share more traits than those within the whole kingdom. One of the characteristics used to classify an animal is whether or not it is symmetrical. Some animals, like sponges, have no symmetry at all. Others, like starfish and jellyfish, have **radial symmetry** where organisms resemble a pie shape. Lastly, animals could have **bilateral symmetry**, meaning that their right and left sides are mirror images of each other, like humans. There are about 35 phyla of animals, but the best-known animal phyla are Porifera, Cnidaria, Platyhelminthes, Nematoda, Annelida, Mollusca, Arthropoda, Echinodermata, and Chordata. Review the characteristics of several animal phyla in the table.

COMMON ANIMAL PHYLA

Phylum	Common Characteristics	Familiar Animals in the Phylum
Porifera	Pores in which water flows into a central cavity	Sponges
Cnidaria	Special stinging structures Radial symmetry and a central body cavity	Sea anemone, coral, jellyfish, hydra
Platyhelminthes	Bilateral symmetry One body opening (mouth)	Tapeworms, planaria, flukes
Nematoda	Bilateral symmetry Two body openings (mouth & anus)	Roundworms, threadworms, Trichinella
Annelida	Segmented Bilateral symmetry	Earthworms, leeches
Mollusca	Soft body Some have shells secreted by a body part called the mantle	Snails, clams, mussels, scallops, oysters, octopus, squid
Arthropoda	Exoskeleton (outer skeleton) Bilateral symmetry Segmented body and jointed appendages	Insects, arachnids (spiders and ticks), crustaceans (lobsters, crabs, shrimp)
Echinodermata	Radial symmetry and an internal calcite skeleton Often covered with spines	Sea stars, starfish, sea urchins, sand dollars, sea cucumbers
Chordata	Nerve cord like the human spinal cord Gill slits at some stage of its life cycle	All vertebrates and some primitive wormlike sea animals

Example 9

A person sorted six animals into these two groups.

Group 1	Group 2
Insect	Fish
Crab	Snail
Spider	Jellyfish

Which characteristic was used for sorting?

A. eyes

B. shell

C. exoskeleton

D. backbone

See page 65 for the answer.

CLASSES OF VERTEBRATES

We are most familiar with animals in the phylum Chordata, like frogs, fish, whales, birds, and humans. The vast majority of animals in this phylum are vertebrates. This means that they have a backbone or spinal column. Vertebrates share other traits as well: They have a mouth and sense organs at one end of their bodies, and during at least one stage of their life cycle they have gill slits. Even humans have gill slits at one point in their development as embryos! Also, all vertebrates have bilateral symmetry.

Within this phylum there are many classes, which further categorize animals into a more specific grouping. Some of the common classes of vertebrates are fish, amphibians, reptiles, birds, and mammals. The organisms in all of the classes share a common ancestor and some similar traits, but there are specific characteristics that distinguish one class from another. Explore the characteristics of the common classes of chordates in the table.

COMMON CLASSES WITHIN PHYLUM CHORDATA

Class of Chordates	Characteristics	Examples
Fish (there are three classes) Jawless fish, Cartilage fish, Bony fish	Jawless fish do not have scales Cartilage fish have jaws and scales Bony fish have a skeleton of bone Cold-blooded	Jawless fish—lampreys, eels Cartilage fish—sharks, rays Bony fish—trout, salmon, tuna
Amphibians	Moist skin Eggs are laid in water Young live in water Adults live on land Cold-blooded	Frogs, toads, salamanders, newts
Reptiles	Have lungs and scales Lay leathery eggs on land Cold-blooded	Turtles, snakes, lizards, alligators
Birds	Feathers and beaks Hollow bones Lay eggs Warm-blooded	Eagle, penguins, geese, sparrow
Mammals	Birth live young Nourish their young with milk Warm-blooded Teeth and hair/fur	Dogs, cats, bears, humans, walrus, otters, whales

Example 10

Which of the following animals belongs in the same class as a dog?

A. snakes

B. salamanders

C. dolphins

D. eagles

See page 65 for the answer.

NAMING ORGANISMS

After a scientist determines the class of animal, they then classify the organism into the smaller groupings of order, family, genus, and finally species. Some common orders of mammals are rodents, carnivora, and primates. The biologic name given to all organisms is the organism's genus and species. A species is a group of organisms that can interbreed successfully. We humans are in the order of primates and in the family hominidae. Our biologic name is *Homo sapiens*. *Homo* is our genus and *sapiens* is our species.

KEY CONCEPT

An organism's biological name is its genus and species.

Example 11

Level of Classification	Human	House Cat	Tiger	Lion
Kingdom	Animalia	Animalia	Animalia	Animalia
Phylum	Chordata	Chordata	Chordata	Chordata
Class	Mammalia	Mammalia	Mammalia	Mammalia
Order	Primates	Carnivora	Carnivora	Carnivora
Family	Hominidae	Felidae	Felidae	Felidae
Genus	*Homo*	*Felis*	*Panthera*	*Panthera*
Species	*sapiens*	*catus*	*tigris*	*leo*

Which two organisms in the chart are the most closely related?

A. human and house cat

B. house cat and tiger

C. tiger and lion

D. house cat and lion

See page 65 for the answer.

ACQUIRED VS. INHERITED CHARACTERISTICS

Organisms are classified based on their characteristics and structural similarities, but what controls these characteristics? Some characteristics are inherited and others are acquired as a result of interactions with the environment. **Inherited traits** are used to classify organisms; they are controlled by genes found in cells and are passed from one generation to the next. **Acquired traits** are not used to classify organisms because they cannot be passed from one generation to next. Ask yourself: If I dye my hair blue, is there any possibility that my children will be born with blue hair? Of course not, because this is an acquired, not an inherited, trait!

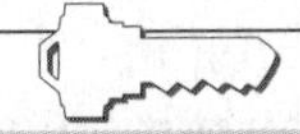

KEY CONCEPT

Some characteristics are inherited, whereas others are acquired as a result of interactions with the environment. Only inherited characteristics can be passed from one generation to the next.

NATURAL SELECTION, EVOLUTION, AND EXTINCTION

Organisms with beneficial traits are more likely to survive and reproduce in nature. For example, in a snowy environment an animal with white fur will have an advantage over an animal with dark fur. The white fur is a beneficial trait for this species. The animal with the white fur will be successful because it is camouflaged from its predators and prey. Because of this, more white-furred animals will make it to their reproductive age and produce offspring with this trait. Eventually, the white-furred animals will dominate the population because they are more adapted to the environment. This is called **natural selection.**

Sometimes there are changes in environmental conditions, such as a rise in temperature or the addition of a new predator. This can affect the survival of a species. Either the population of organisms will gradually adapt to the change or the species will become extinct. Organisms with traits that are helpful in the changed environment are more likely to survive and produce offspring. For example, if the temperature of an environment decreases, rabbits with thicker fur might have a better chance for survival because they can keep warmer than other rabbits with lighter coats. This trait, thicker fur, will be passed down to the next generation of rabbits. Rabbits with lighter fur might not survive as well or reproduce; so this trait will not be passed down to the next generation. This change in the population of rabbits is an example of the process called **evolution**. Evolution is the result of natural selection for beneficial traits. It results in changes in a species and contributes to the diversity of species.

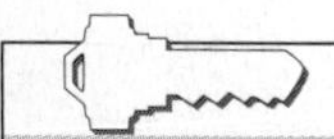

KEY CONCEPT

Organisms with beneficial or helpful traits are more likely to survive changes in environment conditions and produce offspring.

The diversity or variations in traits contribute to a species' chance of survival if the environment changes. Within a species, there are many variations in traits. Just think about the wide range of physical characteristics humans exhibit. We are all of the same species, yet we look so different from one another. Even though there are many variations of traits, in some cases the environment may change and a species may not have beneficial or helpful traits within its population. This can cause the **extinction**, or disappearance, of that species. The fact that we don't have dinosaurs today is the result of extinction. The variety of traits that dinosaurs had did not help them survive when the environment changed. Extinction of species is quite common; actually, the majority of species that have lived on the Earth no longer exist today.

Example 12

A particular environment is becoming warmer over time. A rabbit with which of the following traits would best be able to survive and reproduce in this environment?

A. longer legs

B. larger ears

C. dark-colored, thick fur

D. light-colored, thin fur

See page 65 for the answer.

REPRODUCTION AND HEREDITY

After reading this section you should be able to:

- Describe life cycles of humans and other organisms.
- Describe how the sorting and recombining of genetic material results in the potential for variation among offspring of humans and other species.

LIFE CYCLES

All organisms grow and develop through their life cycles. The life cycle of an organism is the series of changes it goes through from birth to death. These changes occur as it develops and reproduces. The life cycle of a human, for example, lasts for approximately seventy-five years and consists of the following stages: birth, infancy, childhood, adolescence, and adulthood.

Many insects go through four stages of metamorphosis as part of their life cycle: egg, larva, pupa, and adult. The larva is wormlike and, in many species of insects, is called a grub or a caterpillar. The larva builds a cocoon for the next stage, the pupa. During pupation in the chrysalis, or cocoon, the larva is transformed into the adult where new body structures, like wings, emerge. Refer to the figure and follow the stages in the life cycle of an ant.

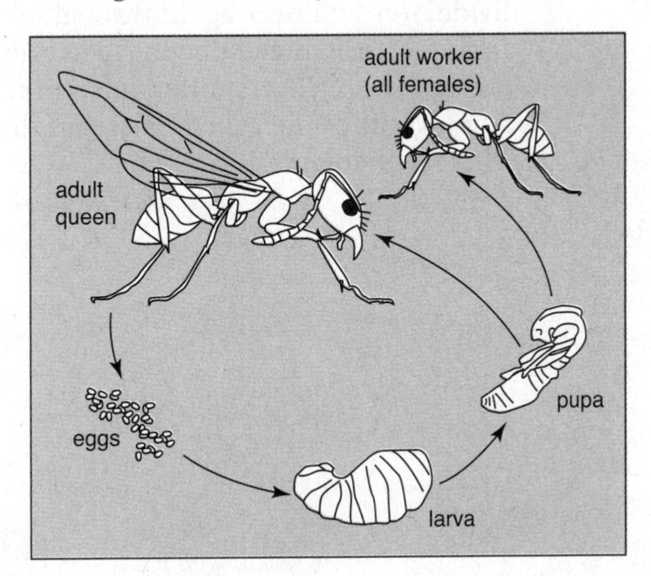

Frogs are another example of an organism that goes through dramatic changes in its lifetime. As with all amphibians, frogs start their lives as an egg in a ball of jelly under water. The young tadpole emerges and has characteristics, like gills and a tail, that enable it to

KEY CONCEPT

All organisms go through unique life cycles. The life cycle of an organism is the series of changes it goes through from birth to death.

survive in the water. Gradually, the young amphibian grows lungs and legs, while losing its gills and tail. It leaves the water to live mostly on land but returns to the water to reproduce, beginning the life cycle again.

Example 13

Which of the following shows the correct order of the life cycle of an insect?

A. adult, pupa, larva, egg

B. pupa, adult, egg, larva

C. egg, pupa, larva, adult

D. egg, larva, pupa, adult

See page 66 for the answer.

GROWTH AND REPRODUCTION

Cells must divide in order for organisms to grow and reproduce. There are two types of cell division, mitosis and meiosis. During mitosis, a cell divides into two new body cells. These new cells are the same as the original cell. This is how some organisms grow; cells divide over and over again through mitosis. On the other hand, during meiosis, a cell divides twice, and this results in four sex cells. The new sex cells are different from the original cell because they contain half of the genetic material than what was in the original cell. (See the figure.)

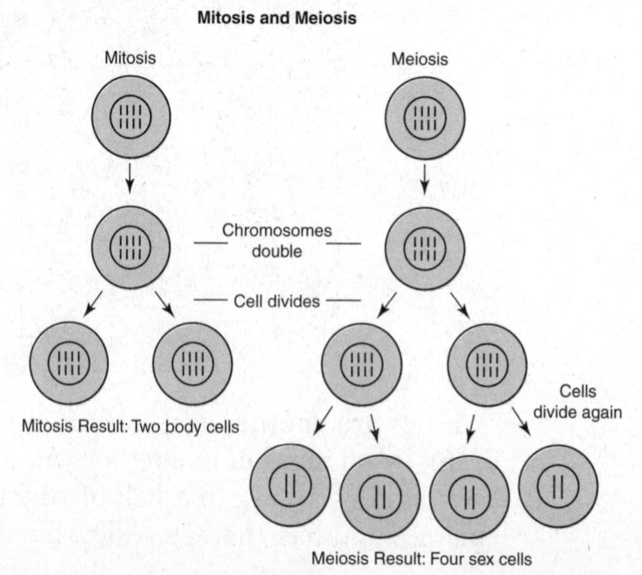

Organisms reproduce at different stages of their life cycle and in different ways. No individual organism can live forever, so reproduction is essential for the continuation of all species. Because all organisms are made of one or more cells, reproduction happens at the cellular level. Cells must divide for reproduction to occur.

Organisms can reproduce in two specific ways, sexually and asexually. In **asexual reproduction**, a single cell divides into two daughter cells through mitosis. The two daughter cells are genetically identical to the parent cell. Organisms like bacteria, algae, flatworms, and some plants reproduce asexually. Therefore, the offspring of these organisms share identical characteristics with their parents.

In **sexual reproduction**, males and females produce sex cells through meiosis. In male animals, the sex cells are called sperm; in male plants, it is called pollen. In females, the sex cell is the egg. The two sex cells, the egg and sperm, combine to create a new organism. The new organism is not identical to either parent; it has a mixture of traits from both. Within species that reproduce sexually, there are a wide variety of traits; sexual reproduction is the main cause of variation in species.

One example of sexual reproduction is seen in flowers. The male structure in a flower is called the stamen, and it produces pollen. During pollination, the pollen attaches to the sticky pistil, the female structure. It travels down the pistil until it reaches the ovary, which contains the female sex cell, the ovule. Fertilization occurs when the pollen cell joins with the ovule and forms the fertilized embryo. This embryo develops into a seed that can eventually grow into a new plant that has characteristics of both parents.

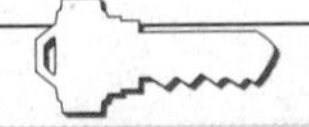

KEY CONCEPT

Some organisms reproduce asexually, resulting in offspring that are genetically identical to the parent. Other organisms reproduce sexually, resulting in offspring that have a blend of characteristics from both parents.

Example 14

Some animals, like sponges, can reproduce both sexually and asexually. How can you test whether a particular sponge organism is the product of sexual or asexual reproduction?

A. Determine whether its genes are identical to the genes of its parent.

B. Determine if its traits are beneficial for survival.

C. Determine if its traits are not beneficial for survival.

D. Determine whether it has acquired characteristics.

See page 66 for the answer.

GENES, CHROMOSOMES, AND DNA

Within the nucleus of every single cell of an organism, there are genetic structures called chromosomes. Chromosomes are made of a substance called deoxyribonucleic acid, or DNA. DNA is a long, twisted, thread-like molecule and it holds the genetic code for building and maintaining an organism. Sections of the DNA correspond to genes, which control a particular trait, like hair color, eye color, or height. (See the figure.)

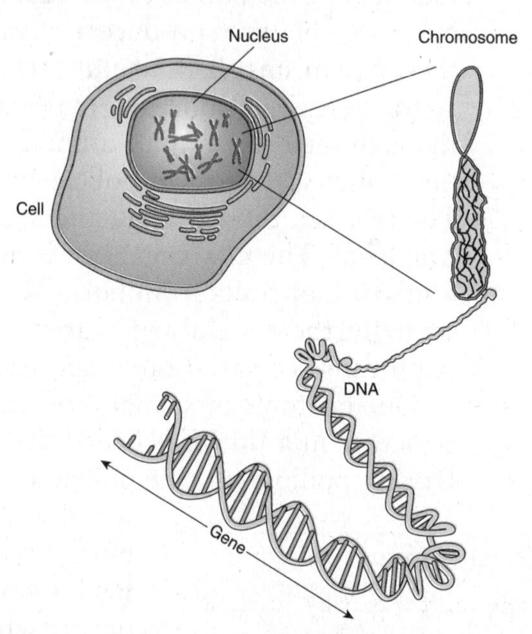

When an organism reproduces asexually, the genetic information in the nucleus of each of its cells is identical to that of the parent. However, organisms produced through sexual reproduction have half of the genetic information from each parent. Different organisms have different numbers of chromosomes in their cells. Humans have 46 chromosomes, fruit flies have eight, and potatoes have 48. During meiosis, sex cells are created that have half the number of chromosomes as a regular body cell. Thus, a human male sperm cell and a female egg have 23 chromosomes each. When they combine to create the fertilized egg, the resulting human offspring has 46 chromosomes in each cell. This results in a blend of characteristics from each parent; therefore, sexually produced offspring are never identical to either of their parents.

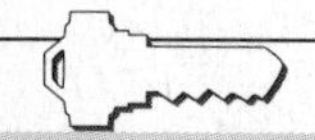

KEY CONCEPT

Genes are located on chromosomes in the nucleus of cells. They carry information that determines inherited characteristics. Sex cells have half as many chromosomes as body cells.

Example 15

The body cell of a butterfly contains 24 chromosomes. How many chromosomes are found in the egg cell of a female butterfly?

A. 24

B. 12

C. 20

D. 48

See page 66 for the answer.

DOMINANT AND RECESSIVE TRAITS

Our genes hold the code to our characteristics. Every organism produced by sexual reproduction has two copies of every gene: one from its mother and one from its father. Because half of the genes come from each parent, we can determine the probability, or chance, of certain traits appearing in their offspring.

Let's look at an example in pea plants. The trait we will examine will be the height of the plant. Pea plants can be either tall or short. The trait is controlled by two alleles, or forms of a gene. Alleles are different versions of a gene. The gene would be height and the alleles would be tall (T) or short (t). Tall (T) is the dominant, or stronger trait, and short (t) is the recessive trait. Recessive traits can be masked or hidden by dominant traits. If a particular pea plant inherits a tall allele from each parent, then it will have the genotype, or genetic makeup, "TT," and phenotype, or physical appearance, of being tall. If another pea plant inherits a short allele from each parent it will have the genotype, "tt," and phenotype, short. We call both of these plants purebred because they have two of the same alleles for a particular trait. However, if a pea plant inherits a tall allele from one parent and a short allele from the other parent, its genotype will be "Tt," and it is called a hybrid; hybrids have two different alleles for a particular trait. The phenotype of this plant will be tall because the tall allele masks the short allele.

A chart called a Punnett square is used to determine the chance of a particular trait appearing in the potential offspring of two organisms. Let's use a Punnett square to determine the potential offspring of a cross between a purebred tall plant and a hybrid pea plant. Follow through the steps in the figure on page 62 to review how to use a Punnett square.

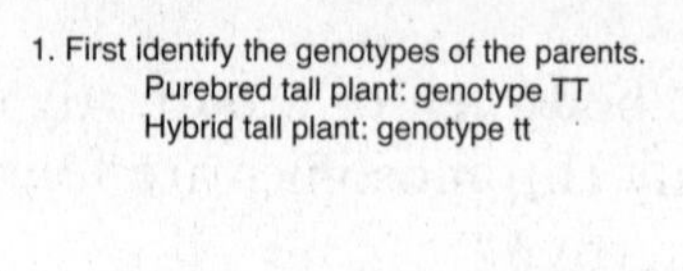

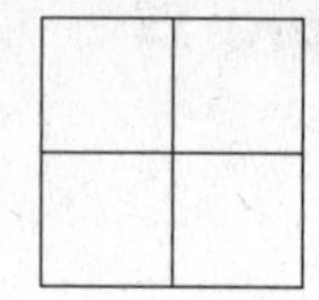

2. Place the genotypes of the parents along the top and side of the chart.

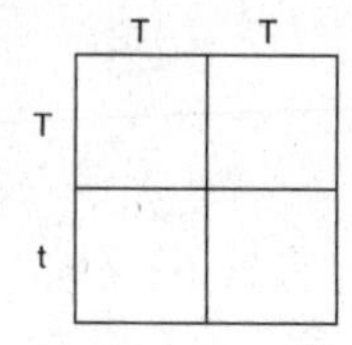

3. The four boxes represent four potential offspring. Transfer the letters down the columns and across the rows.

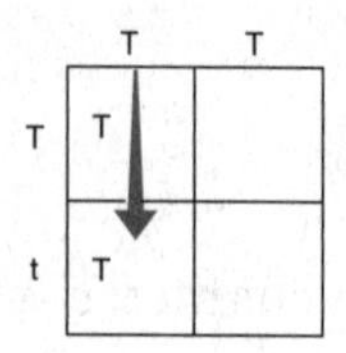

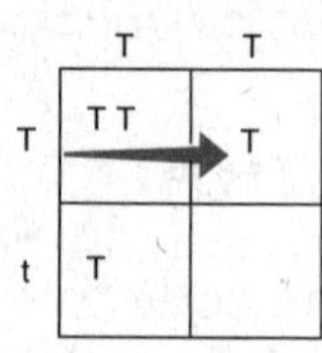

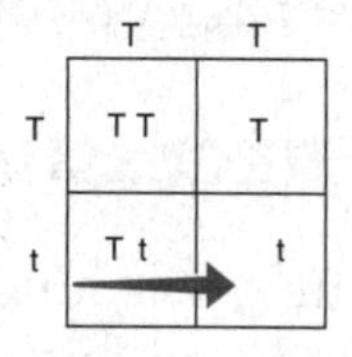

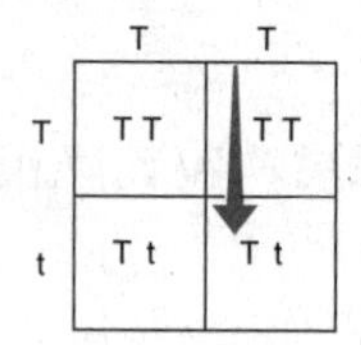

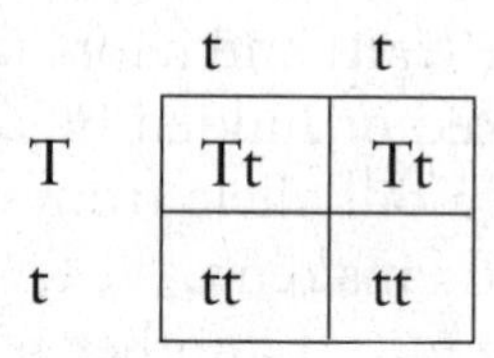

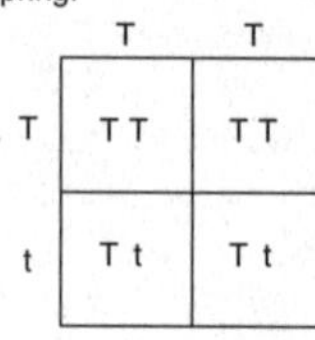

Example 16

	t	t
T	Tt	Tt
t	tt	tt

In pea plants, tall (T) is dominant over short (t). This Punnett square shows a cross between a purebred short plant and a hybrid tall plant. What percentage of the offspring will be tall?

A. 25%

B. 50%

C. 75%

D. 100%

See page 66 for the answer.

SUMMARY

In this chapter, we learned about the diversity, complexity, and interdependence of life on Earth. We also explored how organisms evolve, reproduce, and adapt to their environments.

We began by exploring cells, the basic unit of structure in organisms. Within cells, tiny cell parts work to keep the cell active and functioning. Because most cells are extremely small, we use microscopes to examine them. Using microscopes, scientists observed that many processes occur within cells. The processes of photosynthesis and respiration occur at the cellular level and provide organisms with energy and nutrients.

Multicellular organisms are made of many well-organized cells: Cells join together to form tissue; tissues join together to form organs; and organs work together to form organ systems. Organ systems team up to support bodily functions and maintain the life of the organism.

Because of the vast number and variety of organisms on Earth, scientists classify organisms. The classification system is based on physical structures and moves from broad, general groupings of kingdoms to smaller, specific groupings of genus and species. Also, organisms are named as they are classified; an organism's biological name is its genus and species.

Environments are constantly changing, and this can affect the survival of species. Organisms with inherited characteristics that are beneficial are likely to survive and reproduce. If the organism has traits that will help it to survive, then it will reproduce, and the species will continue existing. However, if no beneficial traits are exhibited in a population of organisms, then the species may become extinct.

All organisms go through unique life cycles where they change and develop. Throughout the life cycle of an organism, cells divide so that the individuals can grow in size. Cell division is also necessary for organisms to reproduce, either sexually or asexually.

If the organism reproduced asexually, the genetic makeup of an organism is the same as its parent. However, if an organism was produced through sexual reproduction, it then has half of its genes from each parent. These genes are found in the nucleus of cells and control the traits of the individual. Some traits are dominant, while others are recessive. The probability of traits appearing in the potential offspring of two individuals can be determined using a Punnett square.

KEY TERMS FROM THIS CHAPTER

organism
cell
cell membrane
nucleus
mitochondria
cytoplasm
vacuole
cell wall
chloroplasts
compound microscope
photosynthesis
respiration
unicellular
multicellular
tissue
organ
organ system
vertebrate
invertebrate
kingdom
phylum
class
order
family
genus
species
prokaryote
eukaryote
autotroph
heterotroph
archaebacteria
eubacteria
protists
fungi
plants
animals
bilateral symmetry
radial symmetry
inherited traits
acquired traits
natural selection
evolution
extinction
life cycle
mitosis
meiosis
asexual reproduction
sexual reproduction
chromosome
DNA
gene
allele
dominant
recessive
genotype
phenotype
purebred
hybrid
Punnett square

ANSWERS TO EXAMPLES

1. **C** The function of the cell nucleus is to control the activities of the cell.

2. **C** Light, carbon dioxide, and water are needed for the process of photosynthesis.

3. **A** Cells are the simplest. Organ systems are the most complex. Similar cells join together to form tissue; tissues join together to form organs; organs work together in systems.

4. **D** The lungs are part of the respiratory system. As you inhale, oxygen from the air is transferred through the lungs to the blood. The blood is part of the circulatory system and delivers the oxygen to cells throughout the body.

5. **A** The esophagus, stomach, and intestines work together as parts of the digestive system.

6. **B** The digestive system works to break down food, and the circulatory system transfers the nutrients from the food to the blood. The blood circulates through the body delivering nutrients to body cells.

7. **B** For biological classification, scientists use the structural similarities to group organisms.

8. **A** Species are the smallest, most specific groupings in the classification system. Organisms of the same species have the most characteristics in common. On the other hand, kingdoms are the largest, most general groupings. Organisms in the same kingdom can be quite different from one another and may share just a few characteristics in common.

9. **C** Insects, spiders, and crabs are all members of the phylum Arthropoda, and they all have exoskeletons. Fish, snails, and jellyfish belong to other phyla of animals.

10. **C** Dogs and dolphins are both mammals. The other animals belong to other classes of chordates. Snakes are reptiles, salamanders are amphibians, and eagles are birds.

11. **C** Tigers and lions are both classified in the same genus. Both humans and house cats are in other genera (plural of *genus*).

12. **D** Light-colored, thin fur would be beneficial in a warm environment because the light color would reflect the sunlight and the thinner texture would be better than thick, heavy fur.

13. **D** The life cycle of insects goes from egg, larva, pupa, to adult.

14. **A** If the genes are identical to its parents, then it was produced through asexual reproduction. If the sponge has some of the genes of its parent, then it was reproduced sexually.

15. **B** Sex cells have half the number of chromosomes as a regular body cell.

16. **B** Two of the four boxes (50%) have the genotype, Tt. This indicates a tall phenotype.

PRACTICE QUIZ

For each of the questions or incomplete statements, choose the best of the answer choices given. Turn to page 236 for the answers.

1. Which of the following organs in a fish can be compared to a human lung?

   A. fins
   B. scales
   C. gills
   D. heart

2. The smallest differences in structure occur in organisms of different

   A. order.
   B. phylum.
   C. class.
   D. species.

3. Which types of characteristics can be inherited?

   A. characteristics that result from exposure to the environment
   B. characteristics that are a result of diet and exercise
   C. characteristics controlled by genes
   D. characteristics produced by accident

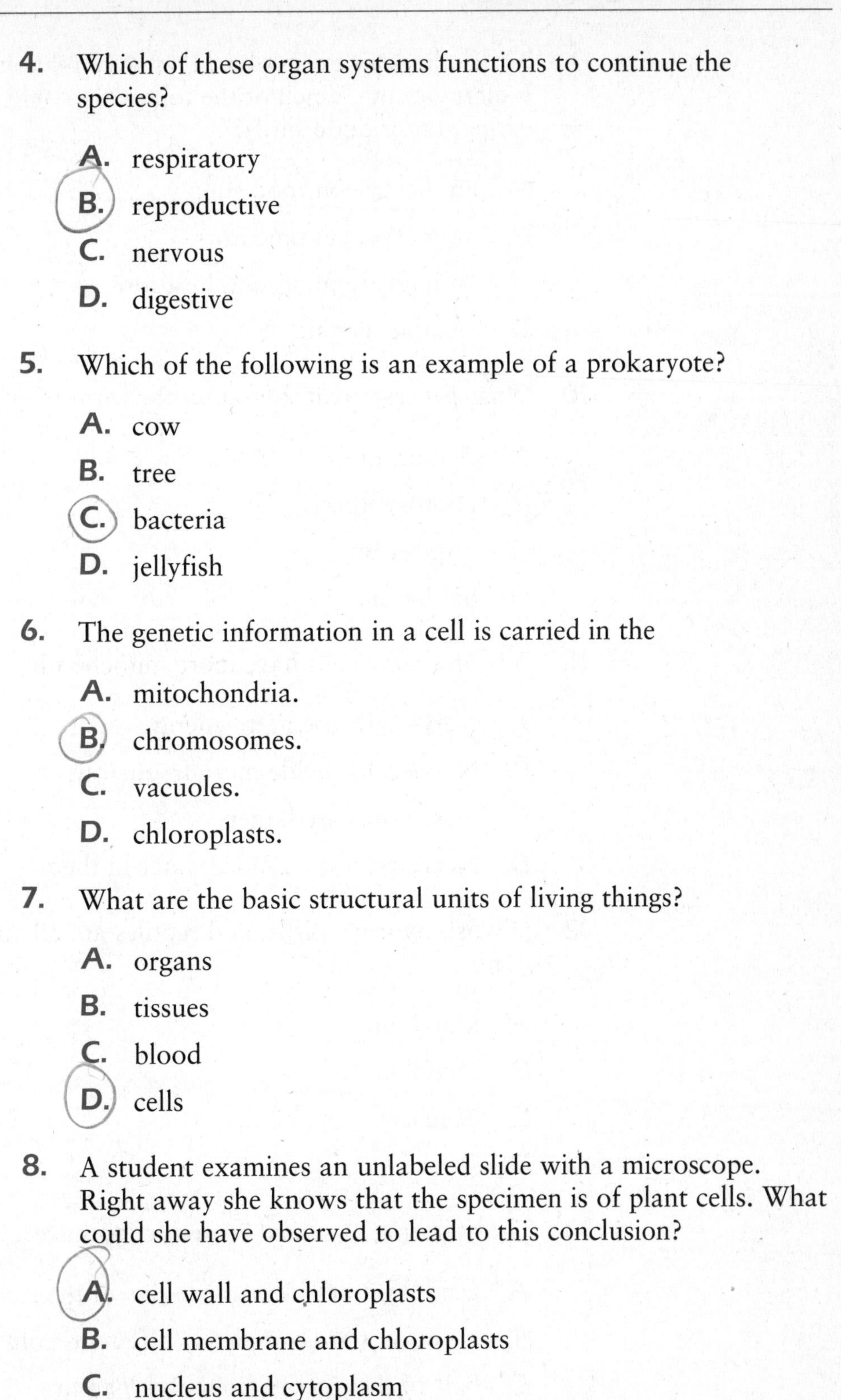

4. Which of these organ systems functions to continue the species?

   A. respiratory
   B. reproductive
   C. nervous
   D. digestive

5. Which of the following is an example of a prokaryote?

   A. cow
   B. tree
   C. bacteria
   D. jellyfish

6. The genetic information in a cell is carried in the

   A. mitochondria.
   B. chromosomes.
   C. vacuoles.
   D. chloroplasts.

7. What are the basic structural units of living things?

   A. organs
   B. tissues
   C. blood
   D. cells

8. A student examines an unlabeled slide with a microscope. Right away she knows that the specimen is of plant cells. What could she have observed to lead to this conclusion?

   A. cell wall and chloroplasts
   B. cell membrane and chloroplasts
   C. nucleus and cytoplasm
   D. cell wall and cytoplasm

9. The dodo is an extinct bird that once inhabited an island in the Indian Ocean. Which of the following could be a cause for the extinction of dodo birds?

   A. an increase in food supply
   B. an increase in predators
   C. an increase in nesting locations
   D. a stable climate

10. What process creates food, in the form of sugar, for plants?

   A. evaporation
   B. photosynthesis
   C. respiration
   D. oxidation

11. Why do nerve cells have more mitochondria than skin cells?

   A. Nerve cells use more energy.
   B. Nerve cells divide more frequently.
   C. Nerve cells are larger.
   D. Nerve cells store more water in them.

12. Jellyfish, worms, birds, and reptiles are all classified in the same

   A. kingdom.
   B. phylum.
   C. genus.
   D. species.

13. The offspring of sexual reproduction have

   A. genes that are identical to one of their parents.
   B. genes that are completely different from their parents.
   C. half of their genes from each parent.
   D. no genes at all.

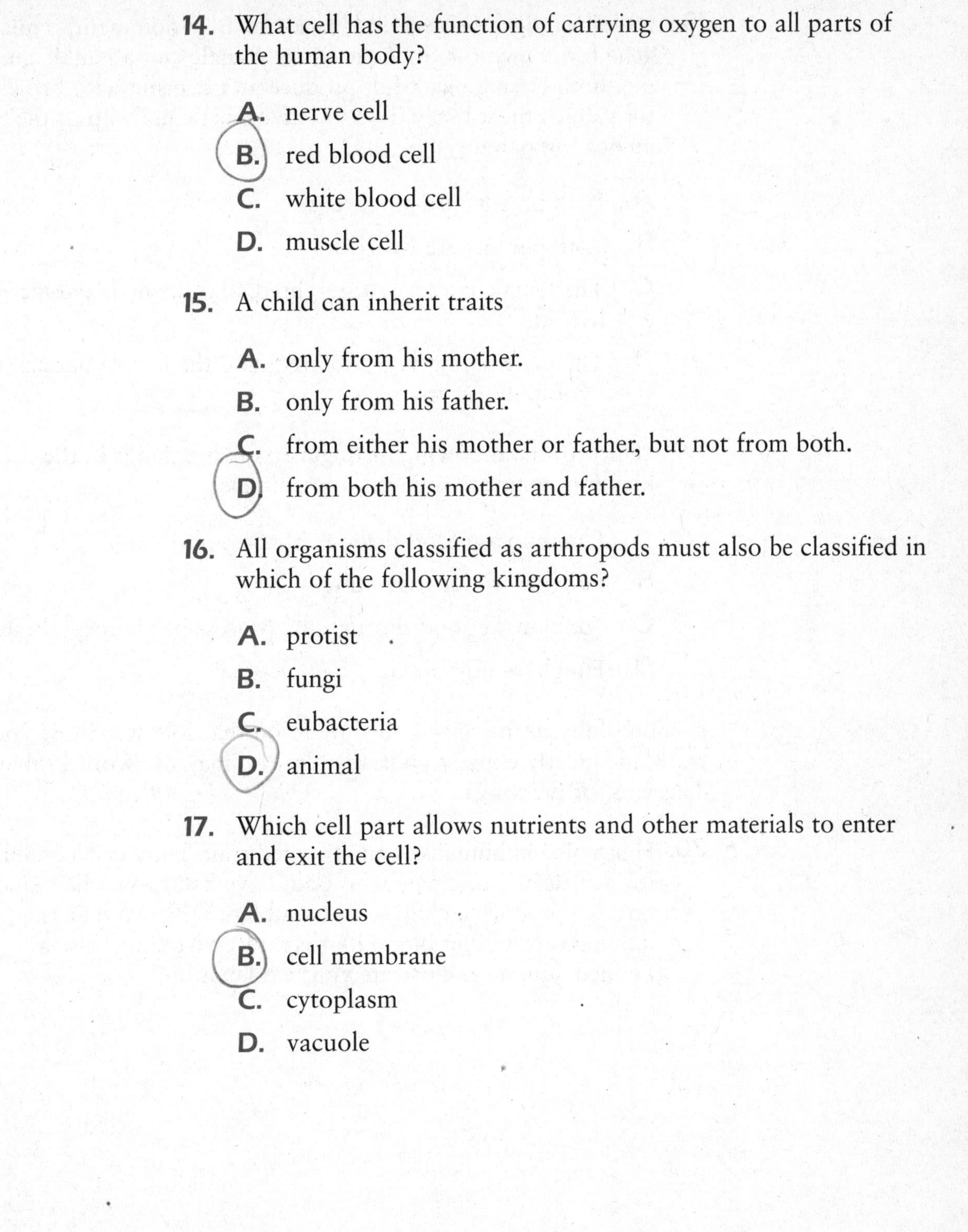

14. What cell has the function of carrying oxygen to all parts of the human body?

A. nerve cell

B. red blood cell

C. white blood cell

D. muscle cell

15. A child can inherit traits

A. only from his mother.

B. only from his father.

C. from either his mother or father, but not from both.

D. from both his mother and father.

16. All organisms classified as arthropods must also be classified in which of the following kingdoms?

A. protist

B. fungi

C. eubacteria

D. animal

17. Which cell part allows nutrients and other materials to enter and exit the cell?

A. nucleus

B. cell membrane

C. cytoplasm

D. vacuole

18. In guinea pigs, the gene for black fur (B) is dominant, while the gene for brown fur (b) is recessive. A male and a female guinea pig, both having black fur, produce an offspring with brown fur. Which most likely describes the genetic makeup of the guinea pig parents?

A. Both parents are purebreds.

B. Both parents are hybrids.

C. The female parent is a purebred, and the male parent is a hybrid.

D. The male parent is a purebred, and the female parent is a hybrid.

19. Which of the following distinguishes living things in the kingdom fungi from plants and animals?

A. Fungi obtain food through absorption.

B. Fungi reproduce sexually.

C. Fungi make food through the process of photosynthesis.

D. Fungi are unicellular.

Respond fully to the open-ended question that follows. Show your work and clearly explain your answer. You may use words, tables, diagrams, or drawings.

20. Hair color in humans is an inherited trait. How is it possible for a male and a female who both have had brown hair since birth to produce a child with blond hair? (Brown hair is a dominant trait, and blond hair is a recessive trait.) Use a Punnett square to illustrate your explanation.

Chapter 3

ENVIRONMENTAL STUDIES

Our world is filled with many kinds of animals, plants, and other organisms. All of these organisms need food, water, and shelter, and they have complex relationships with other organisms and their environment. This chapter will review ecology, the study of the interrelationships between animals and environments, and how humans affect the environment. This chapter is divided into three sections:

- Natural Systems and Interactions: Energy Flow in Ecosystems
  - Ecosystems and Populations
  - Energy Roles and Interactions within Communities
  - Food Chains and Food Webs
  - Succession
- Biomes
  - Land Biomes
  - Water Biomes
- Human Interactions and Impact
  - Catastrophic Natural Events
  - Human Impact
  - Renewable and Nonrenewable Energy Sources

NOTE TO STUDENTS

Thoroughly read each section. As you read pay close attention to the key concepts and try to answer the example questions found in each section. Answers to examples in this chapter are on page 90. At the end of the Chapter 3, assess your knowledge of environmental studies with the practice quiz. Then, check your answers using the answer key provided on page 238.

NATURAL SYSTEMS AND INTERACTIONS: ENERGY FLOW IN ECOSYSTEMS

After reading this section you should be able to:

- Distinguish between the components of an ecosystem.
- Explain how organisms and the environment are interconnected in an ecosystem.

- Describe the relationships that exist between organisms.
- Explain how succession changes the communities in an ecosystem.

ECOSYSTEMS AND POPULATIONS

Organisms and their environments are interconnected; organisms affect the environment, and the environment affects organisms. The **environment** consists of all of the nonliving things around an organism like the light, air, soil, landforms, temperature, water, and climate. These nonliving parts of the environment are called **abiotic factors**. All living things rely on abiotic factors for survival. For example, plants need the water and nutrients in the soil and the carbon dioxide in the air to grow and thrive. Some organisms burrow into the soil or make homes in caves for shelter.

KEY CONCEPT

The environment consists of the nonliving things around an organism.

Organisms depend upon other living things as well as the nonliving things that surround them. An **ecosystem** consists of the living and nonliving things in a particular place that interact with each other. A system is a set of connected parts. If one part of a system changes, it will affect the whole system. This is true of ecosystems as well. For example, in a small aquarium ecosystem, if the temperature of the water increases, the fish within the aquarium may not be able to withstand or survive the higher temperature. Or, in a forest ecosystem, if trees are cut down, many animals will lose their habitats and have difficulty surviving. A **habitat** is the place where an animal or plant lives. A habitat provides shelter and the resources organisms need to live and reproduce. An ecosystem can be as small as a crack in a sidewalk or as large as the ocean; the boundaries of an ecosystem are defined by the ecologist who is studying it.

KEY CONCEPT

An ecosystem consists of the interacting living and nonliving things in a given area.

Example 1

In an aquarium ecosystem, which of the following is an abiotic factor?

A. water

B. algae

C. fish

D. snail

See page 90 for the answer.

Each organism in an ecosystem plays a role. This is called an ecological **niche**, and it relates to where the organism lives and what it does within the ecosystem. The niche of a maple tree in a forest ecosystem includes absorbing sunlight for photosynthesis, providing a habitat for animals, and serving as a source of food for some animals. The niche of bacteria living on a rotting log is breaking down the dead tree into nutrients that will be returned to the soil.

Within an ecosystem are populations of organisms. A **population** consists of a group of organisms of the same species that are found together in a given place and time. The striped bass in Lake Hopatcong are one population. The squirrels in your schoolyard ecosystem also make a population. Because a population consists of organisms of the same species, they can reproduce. A population will increase if an ecosystem can provide abundant resources for a population, and there aren't many diseases or predators. As a population grows, it will require more resources to support all the organisms. The growing populations might use up food resources and locations for habitats. If there is a lack of resources for a population, an increase in predators, or perhaps a harmful change in climate, then there may be a decrease in a population. Populations are dynamic, as they are always changing in response to their environments.

Example 2

Which of the following is an example of a population?

A. mice and rats living in a barn

B. bees in a hive

C. trees, shrubs, and grasses in a field

D. a dog living in a house with a family

See page 90 for the answer.

ENERGY ROLES AND INTERACTIONS WITHIN COMMUNITIES

Populations within an ecosystem interact with one another. They may compete with each other for resources like food, water, and habitats. Or, they may rely on one another for resources. All of the populations in an ecosystem are called a **community**. A community consists of only the living things in an ecosystem.

Example 3

Which of the following sequences shows the levels of ecological organization from simplest to most complex?

A. community, population, ecosystem

B. population, community, ecosystem

C. ecosystem, population, community

D. ecosystem, community, population

See page 90 for the answer.

Energy Roles

Within communities, organisms rely on each other for food and nutrients. There are two broad groupings of organisms based on how they obtain nutrients. An organism is either a producer or a consumer. A **producer** is an organism that makes its own food from nonliving sources. For example, green plants make food, in the form of sugars, in the process of photosynthesis. Algae and some bacteria are also producers.

Consumers cannot make their own food, so they must rely on other organisms as a source of food. Most bacteria and all animals and fungi are consumers. Consumers can be further classified, based on the types of organisms they consume, as an herbivore, carnivore, omnivore, or decomposer. **Herbivores**, like cattle, deer, elephants, and caterpillars, consume plants. They have special adaptations to help them eat plants. For example, some herbivores have teeth that allow them to cut into and grind plant tissue. Some also have complicated stomachs and long intestines that allow for considerable time for the tough plant cells to be broken down and digested. Some herbivores only feed on one type of plant; the koala only feeds on the leaves of eucalyptus trees. However, most herbivores consume a variety of plants. **Carnivores**, like owls, tigers, and hyenas, feed mostly on other animals. Some plants, like the Venus flytrap, are also carnivores, as they entrap and digest insects that land on their leaves. Most carnivorous animals have very strong jaws.

There are some animals that consume both plant and animal matter. These animals are called **omnivores**. Humans, pigs, and chickens are examples of omnivores. Many physical features of omnivores show how they are adapted to consume both plants and animals. Their teeth, for example, may contain both sharp incisors and canine teeth for tearing meat as well as molars for grinding plant matter. Because omnivores can consume a wider variety of food sources, they are usually able to survive in a variety of environments.

Finally, decomposers, like vultures, earthworms, and mushrooms, feed on dead and decaying plants and animals. Decomposers consume and break down dead plant and animal matter until only raw nutrients remain. These nutrients then go back into the environment for other organisms to utilize. This is crucial for the health of an ecosystem because nutrients are always being recycled. Animals that eat dead animals are called scavengers. The dead and decaying animal matter is called carrion. For example, a vulture is a scavenger that will consume carrion such as a rabbit that is dead and decaying in a field.

Example 4

Which of the following is an example of a decomposer at work?

A. a mushroom growing on a log

B. a hawk preying on a mouse

C. a caterpillar eating a leaf

D. an orchid growing in a rainforest

See page 90 for the answer.

Interactions Within a Community

Organisms in ecosystems interact in a variety of ways. First, there is competition between organisms for resources like food, water, and territory. Competition can exist within a population, but it can also exist between organisms of different species. When there are limited resources, competing organisms will die out or adapt.

But animals do not just compete with one another; sometimes they depend on each other for nutrition and survival. Through predation, one animal hunts and consumes another animal. An animal that lives by hunting and feeding on another organism is called a predator, and the animal that the predator feeds on is called the prey. For example, when a tiger stalks, attacks, and feeds on a wild pig, the tiger is the predator, and the wild pig is the prey.

Example 5

Which of these demonstrates predation?

A. a flea living off a dog

B. a cow eating grass

C. an owl hunting a mouse

D. a mosquito biting a human

See page 90 for the answer.

Symbiosis is a relationship in which two organisms live in close vicinity to one another and at least one organism benefits from the relationship. There are three main types of symbiotic relationships—mutualism, commensalism, and parasitism. In **mutualism** both organisms benefit from the relationship. For example, monarch butterflies lay eggs on milkweed plants. In doing so, they pollinate the milkweed flowers. The monarch larvae feed on the milkweed leaves and acquire the poisonous chemicals of the milkweed leaves. Predators are not as likely to eat the larvae and adult butterflies because of the harmful chemicals. Both organisms benefit in this mutualistic relationship between the monarch butterfly and milkweed plant.

In **commensalism**, one organism benefits while the other organism is not affected or harmed. One example of a commensalistic relationship is the cowbird and bison. Cowbirds are usually found by grazing cattle like the bison. When the bison walk through the grass, insects are disturbed and fly up. This makes it easier for the cowbirds to find and eat insects. The bison are not harmed and do not benefit from this relationship.

The most well-known symbiotic relationship is parasitism. In **parasitism**, one organism benefits, while the other organism is harmed. The organism that benefits is called a **parasite**, and the organism that is harmed is called the **host**. Mosquitoes are a common human parasite. The mosquito bites and sucks blood from humans for nutrition. The human is left with an itchy spot where the mosquito bit. The mosquito is the parasite, and the human is the host. Another example of a parasite is the tapeworm. Tapeworms can live in the digestive systems of hosts like humans, dogs, and other vertebrate animals, and absorb food. The host can experience severe stomachaches, diarrhea, and weight loss.

Example 6

Tapeworms can live in the intestines of dogs and absorb nutrients from the food that the dog consumes. The dog can eventually get health problems because of the tapeworm. What relationship exists between the tapeworm and dog?

A. parasitism
B. commensalism
C. mutualism
D. predation

See page 90 for the answer.

FOOD CHAINS AND FOOD WEBS

Energy and nutrients move through an ecosystem from organism to organism. The major source of energy in ecosystems is sunlight. Producers convert the sunlight into chemical energy through photosynthesis. This chemical energy is stored in green plants. When herbivores and other organisms consume plant matter the chemical energy stored within it is transferred to that organism. Energy continues to move through an ecosystem as organisms eat other organisms.

A **food chain** shows a simple transfer of matter and energy through an ecosystem. In a food chain, the arrow represents the flow of energy. Look at the food chain in the figure at the top of page 78. The grasshopper eats grass. Then, the grasshopper is eaten by a mouse, which is in turn eaten by an owl. Eventually, when the owl dies, bacteria will decompose it. Through each link on this food chain, energy and nutrients are transferred to each organism.

In a food chain each level of consumption is called a trophic level. A trophic pyramid shows the trophic levels and the amount of biomass in an ecosystem. (See the figure below.) Biomass is the total mass of all of the organisms in a particular population or in a given area. The producers are the primary producers in an ecosystem and make up the largest amount of biomass. Primary consumers are the herbivores in an ecosystem that consume the producers. Secondary consumers are the animals, omnivores and carnivores, that consume the small herbivores in a food chain. The animals that feed on the secondary consumers are the tertiary consumers, and they represent the least amount of biomass in an ecosystem. In the figure above, grass is the primary producer. The primary consumer is the grasshopper. The mouse is the secondary consumer. And, the tertiary consumer in the food chain is the owl.

Trophic Pyramid

Tertiary Consumers

Secondary Consumers

Primary Consumers

Primary Producers

The direct sequence of organisms in a food chain is far too simple for most ecosystems; a food web displays the complex relationships in an ecosystem in more detailed and accurate manner. A food web shows the interconnected food chains in an ecosystem, and they identify the feeding relationships between producers, consumers, and decomposers. Just as in a food chain, the arrows represent the flow of energy from organism to organism. Photosynthesis provides the foundation of almost all food webs, for it converts sunlight to chemical energy stored in green plants.

Example 7

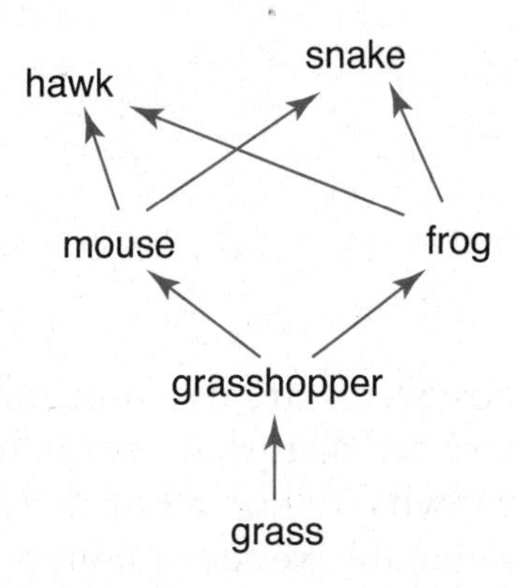

Which of the organisms in this food web is a primary consumer?

A. grass

B. grasshopper

C. hawk

D. snake

See page 90 for the answer.

Take a look at the food web in the figure on page 80. Not only does the food web display what animals eat, but it can also help an ecologist make predictions about an ecosystem. For example, what would happen if there were a decrease in the toad population? Since toads eat grasshoppers, if there were fewer toads, then there would be an increase in the grasshopper population. More grasshoppers would survive since there aren't as many toads preying on them. Also, if there were fewer toads, then there would be more competition within the snake population. Some snakes might even die if they couldn't get enough food.

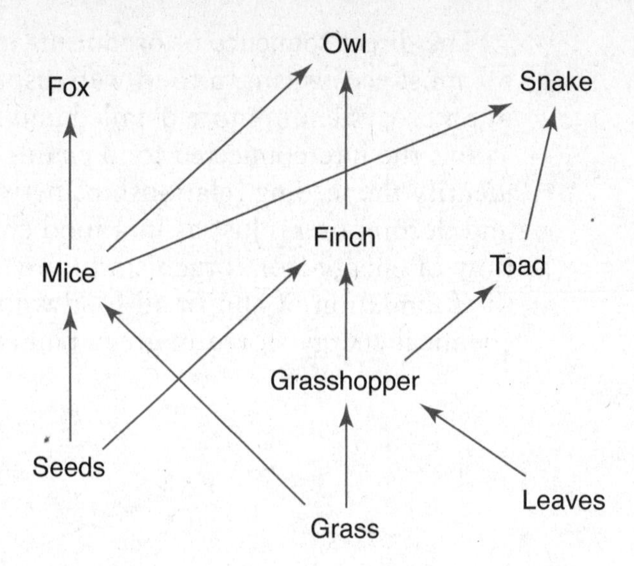

SUCCESSION

Ecosystems are dynamic, always changing over time. Sometimes there are disturbances in an ecosystem, like a glacier, landslide, or fire, which clear all of the vegetation in that area. There are fairly predictable series of communities that will replace one another over time. This is called **succession**. For example, after a glacier retreats from an area, barren land will be exposed. Over time, moss and lichen will begin to grow. Next, as soil begins to develop, grasses and weeds will arrive. Shrubs will eventually grow, followed by pine trees. The final stage of succession is called the **climax community**. For this example ecosystem, a mature oak forest is the climax community.

Lakes also undergo succession. Surprisingly, lakes are rather short-term features of the Earth. Lakes will naturally fill in with sediment and dead plant and animal matter forming peat. Eventually, there will be no standing water, and it will be a marsh with various grasses. In time, the marsh dries up, and shrubs and trees begin to grow. These stages of succession can be spread over hundreds of years.

Example 8

Put the following stages of succession in their proper sequence.

I. Grasses

II. Pine trees

III. Scrub growth

IV. Lichen/moss

A. III, IV, II, I

B. II, I, III, IV

C. I, II, III, IV

D. IV, I, III, II

See page 90 for the answer.

BIOMES

After reading this section you should be able to:

- Distinguish between the land and water biomes.
- Identify the climate and common plants and animals found in each of the major biomes.

The Earth ecosystem is divided into biomes. **Biomes** are large ecosystems that extend over wide regions. They are characterized by dominant plant and animal species as well as by their climate. Scientists have studied and described many biomes. Some of the major land and water biomes are reviewed here.

LAND BIOMES

Coniferous forests have long, cold winters. The summers are short and warm. They receive little precipitation, and most of the precipitation that falls is snow. The trees are evergreens with needle-shaped leaves; this prevents trees' loss of water. Animals like insects, rodents, and squirrels are common in coniferous forests. Bears, wolves, deer, and foxes also live in coniferous forests.

New Jersey is in the **temperate deciduous forest** biome. Think about the plants and animals that are found in our state. The trees in this biome, including oak, maple, elm, and beech, lose their leaves each fall. Many of the animals found in coniferous forests can be found in temperate deciduous forests as well. Bears, deer, fox, hawks, and owls are common. It has a moist climate, with cold winters.

Tropical rainforests occur in areas with high rainfall and high temperatures. The vegetation can be divided into layers. The tall trees with thick straight trunks and leathery leaves extend high into the canopy and emergent layers 20–30 meters off the ground. The other layers are the understory, shrub layer, and forest floor. Different plants and animals are adapted for a particular layer of the rainforest. Tropical rainforests display great **biodiversity**; there are countless species of insects and plants. Monkeys, bats, butterflies, snakes, orchids, and ferns are but a few of the organisms found in tropical rainforests.

Grasslands are biomes where the main vegetation is a variety of grasses. Grasslands can be found in a variety of climates. Tropical grasslands are sometimes called **savannas**; grasslands in North America may be called **prairies**. North American grasslands support animals like bison, deer, rabbits, foxes, owls, and skunks. In other regions, you may find zebras, rhinoceroses, lions, and cheetahs in grasslands.

Dry, often sandy regions with high temperatures and little rainfall are called **deserts**. Animals and plants must be adapted to the arid climate. Cacti are common because they are able to store water. Snakes, lizards, rodents, and termites can be found in deserts.

Tundra is found in the very cold, windy Arctic regions. There is very little precipitation, yet the soil is full of moisture. Little moisture in the soil evaporates since it is usually frozen. There are no trees and the only vegetation consists of a few shrubs, lichens, and mosses. Most of the ground is bare rock or permanently frozen soil, called **permafrost**. Biodiversity is low. Caribou, arctic foxes, and polar bears are a few of the animals that can survive in the tundra.

Example 9

The kangaroo rat is a small rodent that eats only dry seeds. It does not drink water since it gets water from digesting the food it eats. During the hot, dry days, it stays in a burrow and seals the openings to keep in cooler, moister air. In what biome is the kangaroo rat adapted to survive?

A. desert

B. coniferous forest

C. tropical rainforest

D. temperate deciduous forest

See page 90 for the answer.

WATER BIOMES

Wetlands are marshes, swamps, and bogs that form a transitional area between dry land and bodies of water. They are shallow waters with cattails, reeds, and floating flowers. Many bird species, like the heron, are found in wetlands, which provide food and nesting sites for migratory birds. Wetlands are crucial to a healthy environment for they filter water, reduce erosion, and break down pollutants.

Estuaries are located where the freshwater from rivers and streams meets the saltwater from the ocean. The water is brackish, or a mixture of seawater and freshwater. Estuaries are extremely important ecological habitats because they serve as a breeding location for many ocean fish and provide a nursery-like environment for young and growing fish.

The open ocean represents the last and largest biome. Microscopic organisms such as bacteria and algae as well as giant squid and whales coexist in the ocean. The ocean is a varied environment. Even though it covers the majority of the Earth's surface, little is known about many areas of the ocean. The deep-ocean, for example, seems to be diverse in the number of species living there, yet only a small number of individuals appear to exist. The ocean depends primary upon microscopic phytoplankton, a free-floating plant. Even the deep-ocean receives energy from this plant matter as it decays and sinks to the ocean floor.

HUMAN INTERACTIONS AND IMPACT

After reading this section you should be able to:

- Describe the effect of humans on various ecosystems.
- Explain the impact of potential catastrophic events on New Jersey's environment.
- Compare renewable and nonrenewable resources and how they are managed.

Catastrophic natural events and man-made changes affect ecosystems. Humans must understand how natural events alter the environment and how to maintain balance in our ecosystems. As we utilize natural resources and develop neighborhoods and cities, we must attempt to lessen our impact on the environment.

CATASTROPHIC NATURAL EVENTS

When catastrophic natural events occur in New Jersey, human health, the environment, and the economy are impacted. New Jersey is a densely populated state. Communities by the beach are at great risk during storms and **flooding**. Significant damage to homes and businesses will occur if a large storm strikes New Jersey shore communities. This could hurt the economy as homes must be rebuilt and businesses must be reestablished. Scientists and engineers have tried to protect our beaches from potential erosion during storms and flooding, using engineering systems like **jetties**. A jetty is a pier constructed of boulders and concrete that extends from land out into the ocean to protect the shore from storms and erosion. Engineering systems like jetties sometimes cause unexpected changes and harm to the environment.

In addition to erosion, flooding from storms can also have an impact on water quality and cause health concerns. When floodwater enters wells, the well water can become contaminated with pollutants and bacteria. It is important for citizens to be aware of the potential affects of catastrophic events so that they protect themselves.

Although they are a natural part of forest and grassland ecology, **forest fires** can cause vast property damage to homes and businesses. Instead of waiting for a fire to start through natural means, like a lightning strike, ecologists plan and set controlled fires. **Controlled fires** are usually set when winds are low and closely monitored. Through controlled fires, the forest is renewed, and succession begins. This is often performed in the Pine Barrens of New Jersey. Without controlled fires, natural forest fires can become uncontrollable and spread quickly by winds causing widespread damage. Despite these benefits, many people are against controlled fires because many animals may die during the burning and because of other harmful effects to the environment.

HUMAN IMPACT

Human activities have changed the Earth's atmosphere, land, and ocean. Human alterations of the environment have also resulted in a loss in the diversity of plants and animals; some organisms have become endangered and even extinct. The increasing human population puts large demands on the limited resources of our environment.

Our need for energy has also negatively affected the environment. Through burning fossil fuels, harmful gases, like sulfur dioxide, are released into the atmosphere, and other pollutants are discharged into rivers and streams. The sulfur dioxide can cause **acid rain**. Acid rain is precipitation, like rain, snow, or fog, that contains acid from industrial pollution. Acid rain can damage trees and other plants and disturb water ecosystems. Structures like buildings and statues are also damaged by acid rain.

Example 10

Acid rain is mainly caused by

A. solid waste in landfills.

B. gases released from burning fossil fuels.

C. wastewater released into the environment.

D. plowing farmland.

See page 90 for the answer.

Humans have polluted the atmosphere by releasing other chemicals as well. CFCs, **chlorofluorocarbons**, were released into the atmosphere from the 1930s to the 1990s by industries worldwide. This chemical was even found in aerosol hairspray. In the 1970s, scientists discovered that CFCs were destroying ozone gas in the ozone layer of the atmosphere, so they worked to stop the release of this chemical. Scientists even observed a hole in the ozone layer in the atmosphere above Antarctica. The **ozone layer** is found 6 to 30 miles above the Earth's surface and absorbs ultraviolet radiation from the Sun. If CFCs had continued to be released, it may have caused an increase in skin cancer caused by exposure to ultraviolet radiation.

Although the release of CFCs has been halted, humans continue to release greenhouse gases, like carbon dioxide, into the atmosphere. The **greenhouse effect** is a beneficial and natural process in which the atmosphere absorbs the Sun's radiation and keeps the planet warm. Although this process is natural and life on Earth depends on its warming effect, humans are releasing gases that contribute to this effect. Scientists have observed that the overall surface temperatures on Earth are rising; this is called global warming. **Global warming** can cause disastrous effects to many ecosystems, including melting glaciers and ice caps, changing precipitation and drought patterns, destroying habitats, and altering crop productions. Many scientists, politicians, and citizen groups are working to spread the word about global warming and how individuals and industries can reduce greenhouse gas emissions.

Example 11

The burning of fossil fuels releases carbon dioxide into the atmosphere. What is a potential effect of the increase in carbon dioxide gas in our atmosphere?

A. a colder climate

B. a warmer climate

C. an increase in air pressure

D. a decrease in moisture and humidity

See page 90 for the answer.

Other human interventions, like sewers and landfills, all come at a cost to the environment. Sewers transport storm water to streams and rivers. Unfortunately, the storm water can be contaminated by human waste. Then bacteria, like *E. coli*, can disturb stream and river ecosystems. Humans produce tons of garbage and waste. This garbage and other waste is dumped and buried in landfills. Landfills are engineered to have as little impact as possible on the environment by using liners to prevent waste from leaking into the environment and through other safety techniques. As with all technologies, risks are always a part of landfills. Scavengers, like raccoons, can get buried with the garbage and die. Water and soil can be polluted as wastes leak out of the landfill liners. While these interventions benefit people, the trade-off is that the environment may become polluted, and plants and animals may suffer.

RENEWABLE AND NONRENEWABLE ENERGY SOURCES

Humans have caused significant and in some cases irreversible damage to the environment. Because of this, we must try to manage our natural resources in a conservative way. **Renewable resources**, such as water, air, soil, and solar energy, are replenished naturally over the course of time. Even so, humans must manage these resources to ensure that they are not used at a rate higher than the environment's ability to replenish them. Soil, for example, takes a long time to develop. However, it can be easily eroded by wind and water. Steps must be taken to prevent soil from eroding. Grasses and other plants can be planted so that their roots hold the soil in place. The roots also absorb water, which helps prevent the soil from being eroded by water.

Example 12

On flat open fields, farmers plant rows of trees to prevent soil erosion. Which of the following statements best explains how rows of trees can prevent soil from eroding?

A. The trees disrupt wind and reduce erosion from the wind blowing across the land.

B. The trees prove shade for the soil, so the soil stays moist.

C. The trees attract animals that walk across and compress soil.

D. When the trees die, bacteria decompose the plant matter, and the nutrients return to the soil.

See page 90 for the answer.

Nonrenewable resources, like mineral deposits, coal, oil, and natural gas, cannot be restored after use. There is a certain amount of these substances on Earth. When these resources are depleted, the Earth will not be able to replenish the supply; these resources take millions of years to form naturally. Because of this, nonrenewable resources must be conserved and alternative renewable sources of energy, like solar, wind, and geothermal, must be harnessed. Chapter 4 will review how energy is harnessed from these alternative energy sources.

Example 13

Which of the following is a nonrenewable resource?

A. water

B. wind energy

C. solar energy

D. natural gas

See page 90 for the answer.

SUMMARY

In this chapter, we explored ecology, the study of the interrelationships between animals and environments, and how humans affect the environment. Organisms fill a specific niche, or role, in their ecosystem. They compete with other organisms for food, water, and territory. Organisms can be classified by their role within an ecosystem. For example, an organism is either a producer or a consumer, based on whether it produces its own food or consumes other organisms. Food chains and food webs show the relationships between producers and consumers and how energy flows through an ecosystem. Organisms have other complex relationships. Some animals are predators, hunting and eating their prey, while other organisms form symbiotic relationships whereby at least one organism benefits.

Ecosystems are dynamic. Just as the populations of organisms change depending on threats and available resources, the physical environment undergoes changes as well. Through succession, the living things in an ecosystem grow and mature into a climax community.

The Earth is a large ecosystem that can be divided into several biomes. Biomes have a specific climate and characteristic species of plants and animals. Some of the major land biomes are coniferous forest, temperate deciduous forest, tropical rainforest, grassland, desert, and tundra. Some animals and plants are actually adapted to live in several of these biomes. The tropical rainforest is the most diverse with regard to the variety of plants and animals existing in it.

The ocean is the largest biome. While the ocean covers the majority of the Earth's surface, little is yet known about the deep-ocean ecosystem. Estuaries and wetlands provide important habitats for migratory birds and young fish. They are crucial to the health of the environment as they filter water, reduce erosion, and break down pollutants.

The environment is affected by catastrophic natural events and by human activities. Although catastrophic natural events, like storms and fires, are normal, humans work to prevent damage to structures and fragile ecosystems. Humans have caused serious and long-lasting effects on the environment, which have had an impact on many plants and animals. Humans have destroyed many habitats that have, in turn, led to the extinction of many organisms. As we better understand the natural world and its many ecosystems, humans have attempted to lessen their impact on the environment.

We rely on natural resources, both renewable and nonrenewable, for energy, building materials, and more. Through the responsible use of natural resources, humans will have the materials to live and sustain a quality existence for all mankind.

KEY TERMS FROM THIS CHAPTER

ecology
environment
abiotic
ecosystem
habitat
niche
population
community
producer
consumer
herbivore
carnivore
omnivore
decomposer
scavenger
carrion
competition
predation
predator
prey
symbiosis
mutualism
commensalism
parasitism
parasite
host
food chain
trophic level
biomass
primary producer
primary consumer
secondary consumer
tertiary consumer
food web
succession
climax community
biome
coniferous forest
temperate deciduous forest
tropical rainforest
biodiversity
grasslands
savannas
prairies
desert
tundra
permafrost
wetlands
estuaries
brackish
ocean
phytoplankton
flooding
jetties
forest fires
controlled fires
acid rain
chlorofluorocarbons
ozone layer
greenhouse effect
global warming
renewable resources
nonrenewable resources

ANSWERS TO EXAMPLES

1. **A** Water is not a living thing. All of the other responses are biotic, or living parts of an ecosystem.
2. **B** Bees in a hive are a population because they are all of the same species in a given location. All of the other responses include more than one species living together, which by definition is a community.
3. **B** Populations consist of a group of organisms of the same species in a given area. A community consists of all of the populations living in a given area. An ecosystem is the community and the abiotic components of the environment.
4. **A** A mushroom is a decomposer.
5. **C** When an owl hunts a mouse, it is predation. The owl is the predator, and the mouse is prey. Responses A and D demonstrate parasitism. The cow in response B is not "preying" on the grass, it is simply consuming it.
6. **A** The relationship is parasitism. The tapeworm is the parasite, and the dog is the host.
7. **B** A primary consumer consumes producers. The grasshopper consumes grass, which is the primary producer.
8. **D** The correct order of succession is IV lichen/moss, I grasses, III scrub growth, II pine trees.
9. **A** The kangaroo rat is adapted to live in the desert biome. It is adapted for the high temperature and low moisture of deserts.
10. **B** Acid rain is caused primarily by sulfur dioxide gas released when fossil fuels are burned.
11. **B** Carbon dioxide is a greenhouse gas and contributes to the warming of the Earth.
12. **A** A line of trees will obstruct the flow of the wind and reduce soil erosion.
13. **D** Natural gas is a fossil fuel and a nonrenewable resource.

PRACTICE QUIZ

For each of the questions or incomplete statements below, choose the best of the answer choices given. Turn to page 238 for the answers.

1. Which of the following organisms in a forest ecosystem uses energy from the Sun to make its own food?

   A. maple tree
   B. fox
   C. bacteria
   D. grasshopper

2. How are humans classified in our ecosystems?

   A. producers
   B. consumers
   C. decomposers
   D. parasites

3. Which biome contains the most diversity of life?

   A. desert
   B. tropical rainforest
   C. tundra
   D. grassland

4. Green algae can be observed in pond water. What is the role of this organism in the pond ecosystem?

   A. producer
   B. primary consumer
   C. herbivore
   D. decomposer

5. When plants and animals die, what happens to the nutrients and chemical energy stored within the plant and animal matter?

   A. They are returned to the environment by decomposers.

   B. They disappear when the organisms die.

   C. They are absorbed into the atmosphere.

   D. They are transferred to another ecosystem.

6. Grass ⟶ Zebra ⟶ Lion

   Which statement below best describes a relationship shown in the food chain?

   A. The lion is the prey of the zebra.

   B. The zebra is the prey of the grass.

   C. The zebra is the predator of the lion.

   D. The lion is the predator of the zebra.

7. Organisms that absorb nutrients from decaying plants and animals are called

   A. producers.

   B. consumers.

   C. carnivores.

   D. decomposers.

8.

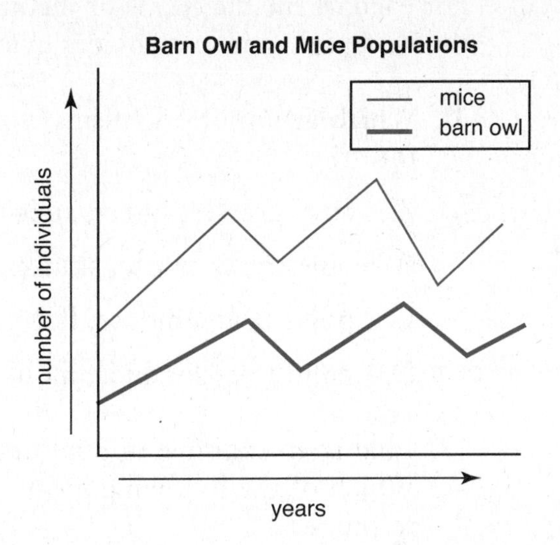

Barn owls prey on mice and other rodents. Which of the following statements is supported by the graph?

A. Barn owls and mice share the same niche.

B. Barn owls are going extinct.

C. The barn owl population increases when the mice population increases.

D. The resources in the barn owl habitat are being depleted.

9. Which sequence shows the correct order in a food chain?

A. producers → carnivores → herbivores

B. carnivores → producers → herbivores

C. producers → herbivores → carnivores

D. herbivores → carnivores → producers

Respond fully to the open-ended question that follows. Show your work and clearly explain your answer. You may use words, tables, diagrams, or drawings. Feel free to use your own scrap paper to write your response.

10. Recent studies indicate that ozone in the upper layers of Earth's atmosphere is being depleted. What caused this depletion, what effect does the depletion of ozone have, and how is this effect harmful to humans?

For each of the questions or incomplete statements below, choose the best of the answer choices given.

11. Which group of organisms would be found in a tropical rain forest?

A. cow, grasses, insects, mice

B. pine trees, moose, skunks, owls

C. polar bears, moose, lichen, moss

D. palm trees, orchids, monkeys, snakes

12. Coal is an example of a natural resource that is not renewable. Which of the following is another example of a nonrenewable resource?

A. oil

B. water

C. soil

D. sunlight

13. Which of the marsupials in the table would be most likely to go extinct if one of its food sources became depleted?

COMMON MARSUPIALS AND THEIR FOOD SOURCES

Marsupial	Food Source
Koala	Eucalyptus leaves
Kangaroo	A variety of grasses and roots
Wombat	Grasses, tree bark, shrub roots
Bandicoot	Insects, earthworms, seeds

A. koala

B. kangaroo

C. wombat

D. bandicoot

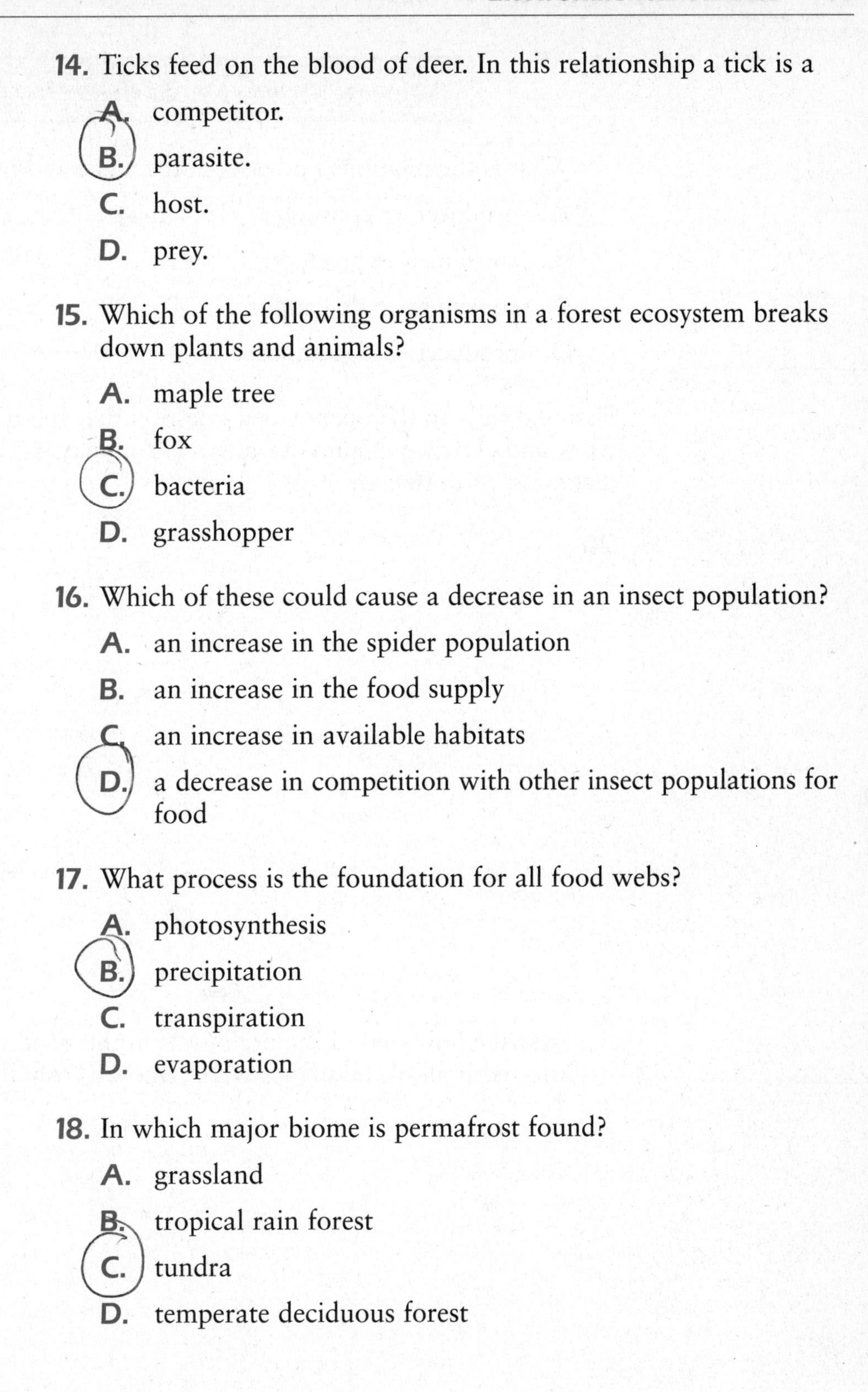

14. Ticks feed on the blood of deer. In this relationship a tick is a

A. competitor.
B. parasite.
C. host.
D. prey.

15. Which of the following organisms in a forest ecosystem breaks down plants and animals?

A. maple tree
B. fox
C. bacteria
D. grasshopper

16. Which of these could cause a decrease in an insect population?

A. an increase in the spider population
B. an increase in the food supply
C. an increase in available habitats
D. a decrease in competition with other insect populations for food

17. What process is the foundation for all food webs?

A. photosynthesis
B. precipitation
C. transpiration
D. evaporation

18. In which major biome is permafrost found?

A. grassland
B. tropical rain forest
C. tundra
D. temperate deciduous forest

19.

Algae ⟶ Minnow ⟶ Bass ⟶ Human

What is the relationship between the algae and minnow?

A. producer → consumer
B. consumer → producer
C. consumer → decomposer
D. producer → decomposer

Respond fully to the open-ended question that follows. Show your work and clearly explain your answer. You may use words, tables, diagrams, or drawings.

20.

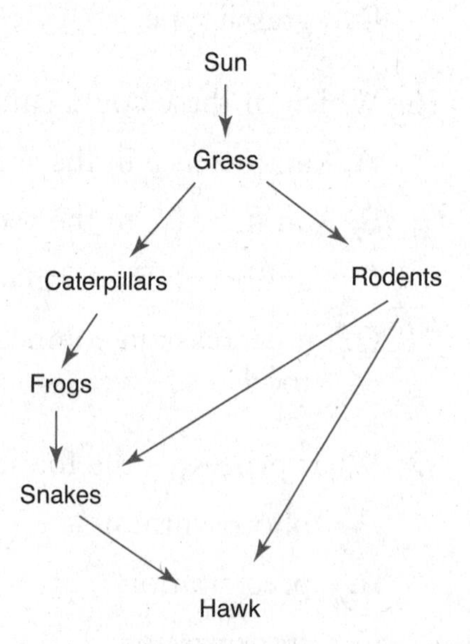

Describe how two of the organisms in this food web would be affected if all of the snakes were removed from the ecosystem.

Chapter 4

PHYSICAL SCIENCE

Physical science is the study of our physical or nonliving world. It involves chemistry, energy, forces, and motion. In order to understand the physical universe, it is necessary to have a strong understanding of the nature of matter and energy. First, we will review basic principles of matter. Then, we will review interactions between energy and matter and examine the physical principles of motion that operate in our universe. Finally, we will review the laws that govern how and why objects move. This chapter is divided into three sections:

- The Structure and Properties of Matter
  - Physical Properties
  - Elements
  - Periodic Table of Elements
  - Metals, Nonmetals, and Noble Gases
  - Atoms and Molecules
  - Phases of Matter
  - Mixtures
  - Physical and Chemical Changes
- Energy
  - Forms of Energy
  - Energy Transformations
  - The Transfer of Heat Energy
  - Movement of Energy in Waves
  - The Sun and the Electromagnetic Spectrum
  - Light Energy
- Motion and Forcers
  - States of Energy
  - Forces and Friction
  - Gravity
  - Newton's Laws of Motion

NOTE TO STUDENTS

Thoroughly read each section. As you read, pay close attention to the key concepts and try to answer the example questions found in each section. Answers to examples in this chapter are on page 132. At the end of Chapter 4, assess your knowledge of physical science with the practice quiz. Then, check your answers using the answer key provided on page 240.

THE STRUCTURE AND PROPERTIES OF MATTER

After reading this section, you should be able to:

- Recognize common physical and chemical properties of matter.
- Identify the groups of elements that have similar properties, including metals, nonmetals, and noble gases.
- Know the arrangement and motion of particles in the phases of matter.
- Describe the properties of mixtures and techniques used for separation.
- Recognize evidence of a chemical change.

PHYSICAL PROPERTIES

Everything around us is matter. **Matter** is anything that takes up space and has mass. You are made of matter; even the air you are breathing is made of matter. All of the stuff around us is made of different types of matter. We can learn more about a sample of matter by examining its physical properties. **Physical properties** can be measured and observed; they can include the boiling point, melting point, and density of a substance. Physical properties don't change depending on the size of the sample of matter you may have. For example, whether you have 10 milliliters of water or 100 milliliters of water, the **boiling point**, or the temperature at which a particular liquid boils at a fixed pressure, will always be 100 degrees Celsius. Not all types of matter have the same boiling point, though. For example, the boiling point of ethanol, a type of alcohol, is 78 degrees Celsius. Another physical property is melting point. The **melting point** is the temperature at which a solid will turn into a liquid. The melting point of water is 0 degrees Celsius, but the melting point of iron is 1,535 degrees Celsius!

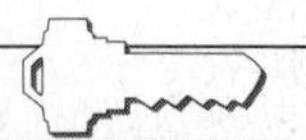

KEY CONCEPT

Physical properties don't change depending on the size of the sample of matter you may have.

Density is yet another physical property of matter. Density is the amount of mass per unit volume of an object. In other words, it is a measurement of how closely packed together the matter is. To calculate the density of an object, divide its mass by its volume.

$$\text{Density} = \frac{\text{Mass}}{\text{Volume}}$$

Mass is the amount of matter in an object, and it can be measured using a triple-beam-balance. **Volume** is the amount of space an object takes up, and it can be measured as described in "Mathematical Applications" in Chapter 1. When a piece of wood is placed in water it floats because it is less dense than the water. If a penny is dropped in water, it sinks because it has a higher density than water. A helium balloon floats because helium gas is less dense than air.

Example 1

$$\text{Density} = \frac{\text{Mass}}{\text{Volume}}$$

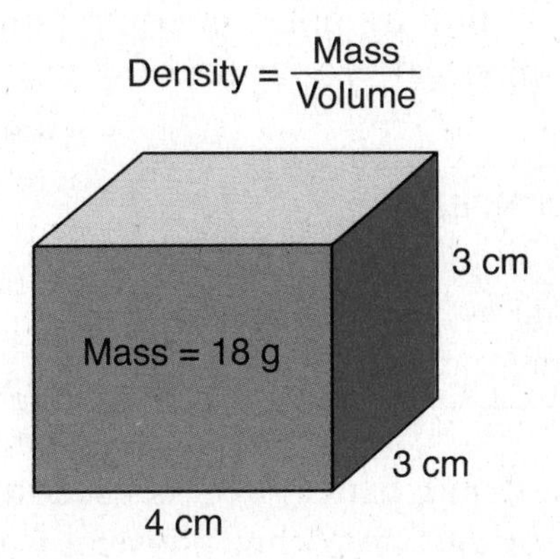

What is the density of the object in this picture?

A. 3 g/cm^3

B. 2 g/cm^3

C. 1 g/cm^3

D. 0.5 g/cm^3

See page 132 for the answer.

Equal volumes of different substances usually have different masses; this is because different substances have different densities. (See the top left figure on page 100.) For example, if you placed one cubic centimeter of gold and one cubic centimeter of silver on a balance, the gold side of the balance would be weighed down because gold is denser than silver. Even though both samples have the same volume, they have different masses due to the density of each samples. (See the top right figure on page 100.)

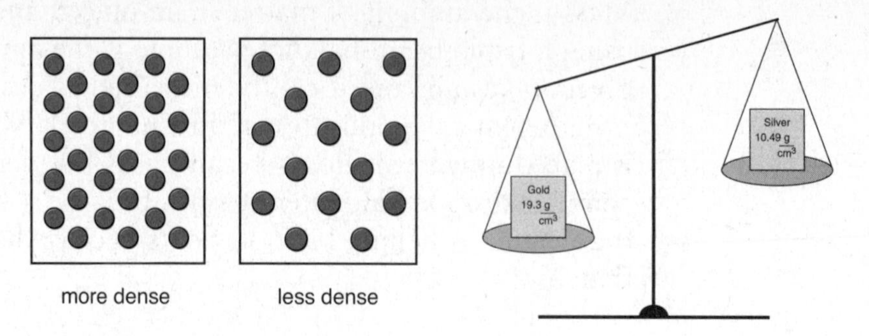

Boiling point, melting point, and density are only a few of the physical properties of matter. Physical properties not only help scientists identify matter but also help them determine potential uses for it. Conductivity, or the ability to transmit heat, electricity, or sound, is another physical property. Metals are good conductors of electricity; therefore, copper (a metal) is used for electrical cables. Iron has the physical property of being very strong; hence, iron is used in the construction of buildings and bridges. Another physical property of iron is its magnetic property.

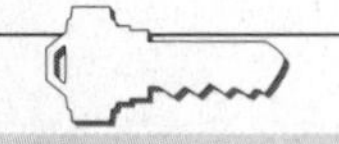
KEY CONCEPT

Equal volumes of different substances usually have different masses; this is because different substances have different densities.

ELEMENTS

Matter is made up of various materials. Every substance we observe is either an element or a compound. An element is a simple substance that has characteristic physical and chemical properties. A compound is a substance made of two or more elements chemically combined. Most of the materials around us are compounds. More than 100 different elements have been identified, but only around 88 elements naturally occur on Earth. Most materials on Earth are made of a few of the most common elements. Oxygen, silicon, aluminum, and iron are a few of the most common elements found on Earth.

Atoms are the fundamental particles of an element. They have all of the chemical and physical properties of the element. They are incredibly small and consist of protons, neutrons, and electrons. In the center of an atom is the nucleus. The nucleus consists of positively charged protons, and neutrons, which have no charge. The negatively charged electrons spin around the nucleus. An element is a substance composed of one type of atom. For example, the element hydrogen has one

KEY CONCEPT

Atoms are the fundamental particles of an element and have all of the chemical and physical properties of that element.

proton and one electron. If it were to have one more proton, then it would be a different element. The element helium has two protons and two electrons. (See the figure.)

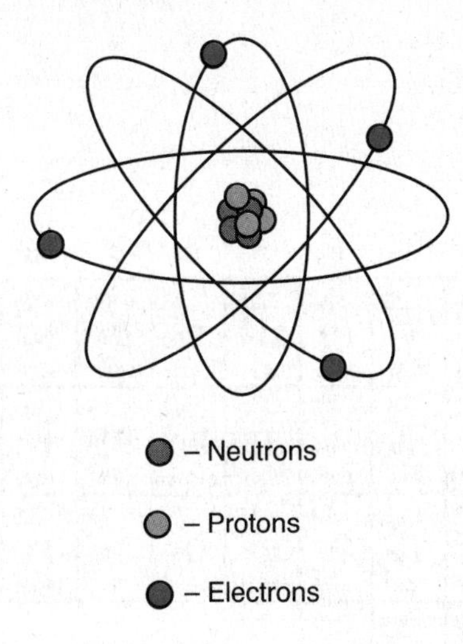

Example 2

Which of the choices listed below is the smallest quantity of an element?

A. mixture

B. atom

C. molecule

D. compound

See page 132 for the answer.

PERIODIC TABLE OF ELEMENTS

The elements have been organized into a table called the Periodic Table of Elements. This table is very useful because it helps scientists understand the interactions between elements, the building blocks of matter. (See the figure on page 102.)

Periodic Table of the Elements

1	2	3	4	5	6	7	8	9	10	11	12	13	14	15	16	17	18
2 H 1.01																	2 He 4.00
3 Li 6.94	4 Be 9.01											5 B 10.81	6 C 12.01	7 N 14.01	8 O 16.00	9 F 19.00	10 Ne 20.18
11 Na 22.99	12 Mg 24.30											13 Al 26.98	14 Si 28.09	15 P 30.97	16 S 32.07	17 Cl 35.45	18 Ar 39.95
19 K 30.10	20 Ca 40.08	21 Sc 44.96	22 Ti 47.88	23 V 50.94	24 Cr 52.00	25 Mn 54.94	26 Fe 55.85	27 Co 58.93	28 Ni 58.69	29 Cu 63.55	30 Zn 65.39	31 Ga 69.72	32 Ge 72.61	33 As 74.92	34 Se 78.96	35 Br 79.90	36 Kr 83.80
37 Rb 85.47	38 Sr 87.62	39 Y 88.91	40 Zr 91.22	41 Nb 92.91	42 Mo 95.94	43 Tc (97.91)	44 Ru 101.07	45 Rh 102.91	46 Pd 106.42	47 Ag 107.87	48 Cd 112.41	49 In 114.82	50 Sn 118.71	51 Sb 121.75	52 Te 127.60	53 I 126.90	54 Xe 131.29
55 Cs 132.91	56 Ba 137.33	57 La 138.91	72 Hf 178.49	73 Ta 180.95	74 W 183.85	75 Re 186.21	76 Os 190.23	77 Ir 192.22	78 Pt 195.08	79 Au 196.97	80 Hg 200.59	81 Tl 204.38	82 Pb 207.2	83 Bi 208.98	84 Po (208.98)	85 At (209.99)	86 Rn (222.02)
87 Fr (223.02)	88 Ra (226.03)	89 Ac (227.03)	104 Rf (261.11)	105 Ha (262.11)	106 Sg (263.12)												

58 Ce 140.12	59 Pr 140.91	60 Nd 144.24	61 Pm (144.91)	62 Sm 150.36	63 Eu 151.97	64 Gd 157.25	65 Tb 158.93	66 Dy 162.50	67 Ho 164.93	68 Er 167.26	69 Tm 168.93	70 Yb 173.04	71 Lu 174.97
90 Th 232.04	91 Pa 231.04	92 U 238.03	93 Np (237.05)	94 Pu (244.06)	95 Am (243.06)	96 Cm (247.07)	97 Bk (247.07)	98 Cf (251.08)	99 Es (252.08)	100 Fm (257.10)	101 Md (258.10)	102 No (259.10)	103 Lr (262.11)

At first glance, the Periodic Table of Elements looks like a bunch of boxes with numbers and letters in them. However, this table is highly organized and very helpful. The columns on the Periodic Table are known as groups, and the rows are known as periods. Each box represents an element. The elements are arranged by their atomic numbers. The atomic number of an element is the number of protons in an atom of that element. For example, the element oxygen has an atomic number of 8, and therefore an atom of oxygen will have eight protons. Also, in the box for each element is a chemical symbol and its atomic weight. A chemical symbol is a one- or two-letter symbol for an element. The chemical symbol for the element oxygen is "O." The atomic weight is a bit more complicated. (See the figure.)

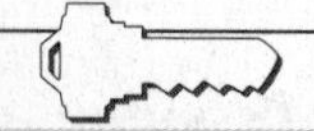

KEY CONCEPT

On the Periodic Table of Elements, the elements are arranged by their atomic numbers. The atomic number of an element is the number of protons in an atom of that element.

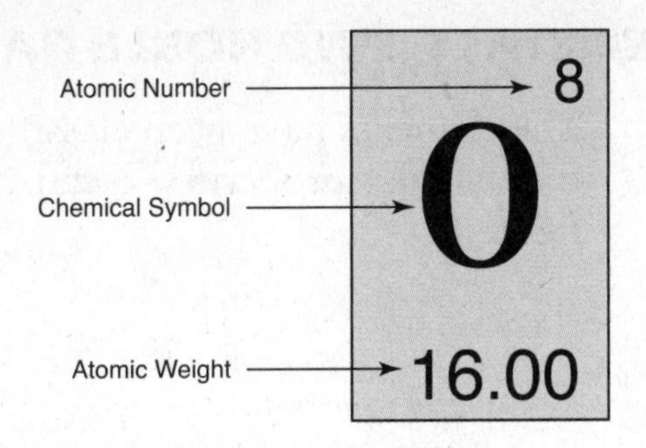

Example 3

Using the Periodic Table, which of the following elements has ten protons?

A. hydrogen

B. lithium

C. helium

D. neon

See page 132 for the answer.

The atomic weight can be used to determine the number of protons and neutrons in the nucleus of an atom. Let's say we want to determine the number of protons, neutrons, and electrons in an atom of oxygen. The number of protons equals the atomic number. For sodium, there are 11 protons and 13 neutrons in the nucleus of its atoms, and 11 electrons circling around the nucleus. (See the figure.)

For any element...
Number of Protons = Atomic Number
Number of Electrons = Number of Protons = Atomic Number
Number of Neutrons = Mass Number – Atomic Number

For Sodium...
Number of Protons = 11
Number of Electrons = 11
Number of Neutrons 23 – 11 = 12

METALS, NONMETALS, AND NOBLE GASES

The elements have been classified into three categories based on their similar properties: metals, nonmetals, and noble gases. (See the figure.)

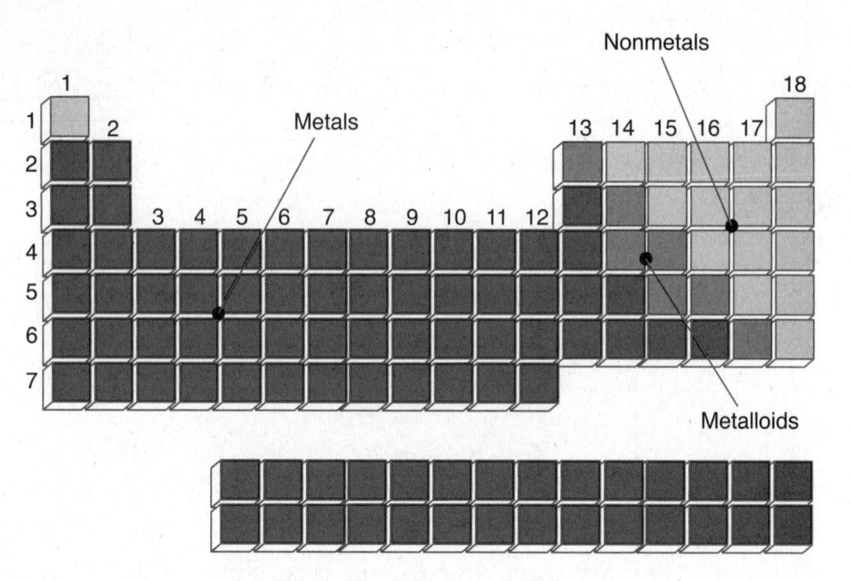

Metals have the following physical properties: conductivity, luster, malleability, and ductility. They are good conductors of heat and electricity, and they have luster, which means they have a glossy appearance. Metals are malleable, meaning they can be shaped and formed by hammering and through pressure. Finally, they can be drawn or made into long, thin wires; this is called the property of being ductile. Gold, tin, aluminum, copper, and silver are examples of metals.

> **KEY CONCEPT**
> Metals have the following physical properties: conductivity, luster, malleability, and ductility.

Nonmetals are poor conductors of heat and electricity. They have a dull appearance, and they are neither malleable nor ductile. There are some elements that have characteristics of both metals and nonmetals; these elements are called metalloids.

The elements in group 18, or the last column on the right side of the Periodic Table of Elements, are called the noble gases. Helium, neon, and argon are examples of noble gases. These elements do not normally react with other elements. They have little or no chemical activity.

Example 4

A student observing a substance recorded her results in the following table.

Property	Observations of Sample
Heat conduction	Good
Melting point	High
Density	$10.5 g/cm^3$
Luster	Shiny
Malleable	Yes
Ductile	Yes

Using the results in the table, what conclusion can you reach about the material?

A. The substance is a metal.

B. The substance is a not a metal.

C. The substance is a noble gas.

D. The substance is a compound.

See page 132 for the answer.

ATOMS AND MOLECULES

All matter is made of atoms. Atoms join together to form molecules of various compounds. A molecule is composed of two or more atoms. Some molecules, like oxygen gas, are very simple. An oxygen gas molecule consists of two oxygen atoms bonded together. Other molecules are more complex. For example, sugar, or sucrose, consists of twelve carbon atoms, twenty-two hydrogen atoms, and eleven oxygen atoms.

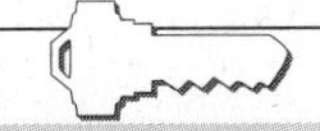

KEY CONCEPT

Atoms join together to form molecules of various compounds. A molecule is composed of two or more atoms.

Symbols are used to represent atoms. The chemical symbol for each element is found in the Periodic Table of Elements. H is the chemical symbol for hydrogen, and O is the chemical symbol for oxygen. A chemical formula is used to represent molecules. For example, the chemical formula for a water molecule is H_2O. Refer to the figure of the water molecule on page 106, which consists of two hydrogen atoms and one oxygen atom. We know this because the number to the bottom right of the chemical symbol indicates how many atoms of that element are found in the molecule. If there is no number to the bottom right of the chemical symbol (as it is for oxygen in the chemical formula for water), then it is assumed that there is only one atom of the element in that molecule. The number 2 beside the chemical symbol for hydrogen indicates that there are two atoms of hydrogen in a molecule of water. Since there is no number to the bottom right of the chemical symbol for oxygen, then we know that there is only one atom of oxygen in a molecule of water.

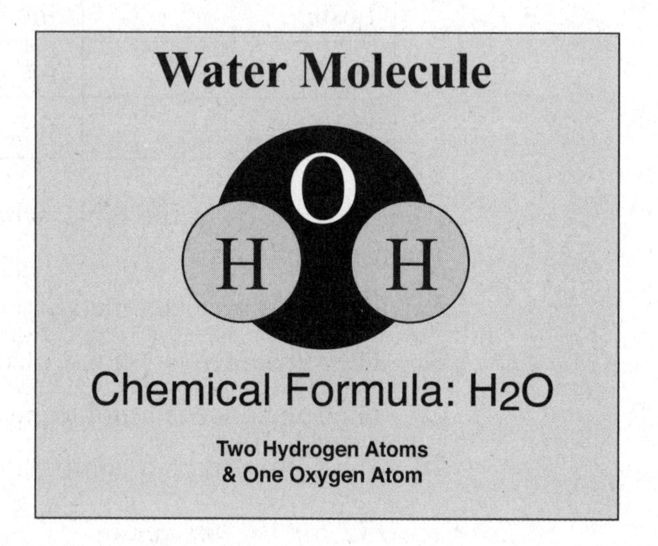

Example 5

Compounds are made up of more than one element. Which of the following represents a compound?

A. O_2

B. H_2O

C. He

D. O_3

See page 132 for the answer.

PHASES OF MATTER

Matter can exist in various forms and phases. There are three basic phases of matter: solid, liquid, and gas. A solid has a definite shape and a definite volume. A liquid has a definite volume, but it will take the shape are whatever container it is in. A gas has no definite volume and no definite shape; the atoms or molecules of a gas will diffuse or spread out to fill whatever container it is in. The phase of matter is determined by the arrangement and motion of atoms and molecules. Atoms and molecules are always in motion because these particles have energy. In a solid, the molecules have less energy and move slower. Because of this, the particles are attracted to one another and arrange themselves closer together and in a particular formation. The molecules in a liquid have more energy than when they are in the solid phase. The molecules move faster and are randomly arranged. In a gas, the molecules move quite fast, exert little or no attraction to one another, and move around in a random fashion. The motion and distance between particles determines the phase of matter. (See the figure.)

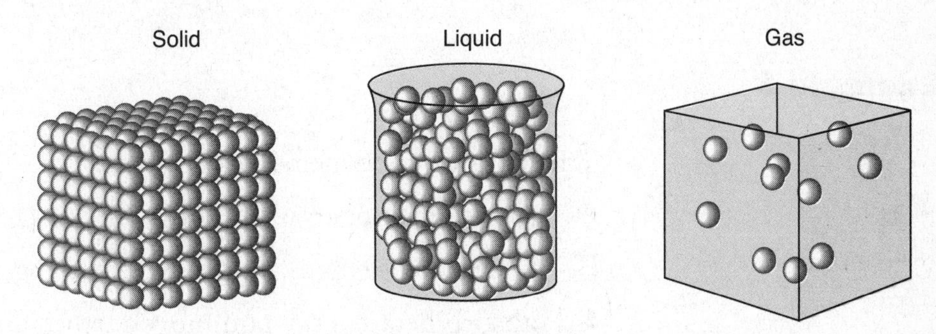

Matter can change from one phase to another. The change is usually due to a change in pressure or the absorption or release of heat energy. For example, water in the solid state is called ice. If the ice cube is heated, then the water molecules will start to move faster and farther apart. The ice will melt into the liquid phase of water. This is called the melting point. If water is heated further, then the liquid will boil and change to the gas phase, water vapor. This is called the boiling point. (See the figure on page 108.)

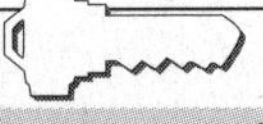

KEY CONCEPT

The phase of matter—solid, liquid, or gas—is determined by the arrangement and motion of atoms and molecules.

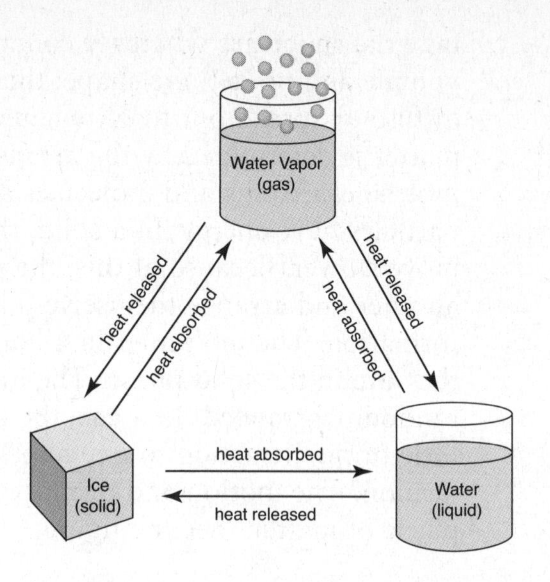

Example 6

When an ice cube melts, its molecules

A. release heat energy and move farther apart.

B. release heat energy and move close together.

C. absorb heat energy and move farther apart.

D. absorb heat energy and move close together.

See page 132 for the answer.

MIXTURES

Many of the substances we encounter are mixtures. A **mixture** is a combination of two or more substances in varying amounts that are physically combined. A mixture can usually be separated into its original substances using one or more of their characteristic properties. For example, if you have a mixture of sand and iron filings (small pieces of iron), you could use a magnet to separate the iron filings from the sand. This is possible because iron has the physical property of magnetism, but sand is not magnetic.

A **solution** is another type of mixture. Some materials, like sugar and salt, are soluble in water. **Soluble** means that the material can be dissolved in a solution. If sugar is dissolved in water, the sugar is called the **solute,** and the water is called the **solvent**. The solute is the substance that is dissolved. The solvent is the material that the solute is dissolved in. Usually the solvent is water or another liquid.

Even a solution of water and sugar can be separated. Because the sugar is dissolved in the water, the sugar cannot be filtered or screened out. But, if the sugar water is boiled or if you wait until all of the water evaporates, you will be left with the sugar.

Some mixtures are more complex and take several steps to separate. How would you separate a mixture of salt water, small wood particles (like saw dust), sand, and iron filings? There are a few ways to accomplish this task. One way is to first skim the wood particles off the surface of the water. Since wood is less dense than water, it will float on the surface. Then, pour the salt water with sand and iron filings through filter paper in a funnel. Next, boil the salt water or allow the water to evaporate so that you are left with the salt crystals. Lastly, use a magnet to separate the iron filings from the sand. In this process, we used the characteristic properties of each substance to separate the mixture.

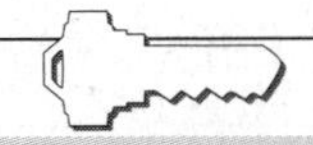

KEY CONCEPT

A mixture can usually be separated into its original substances using one or more of their characteristic properties.

Example 7

Which of the following best shows particles of a mixture?

A

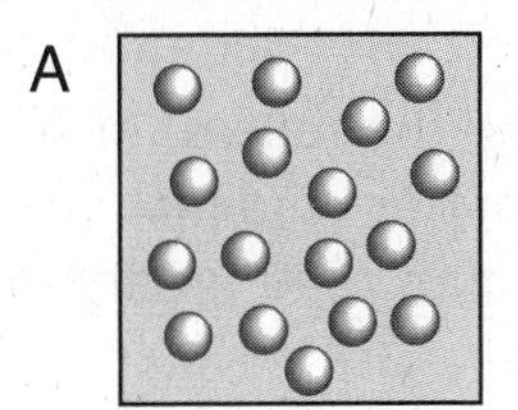

B

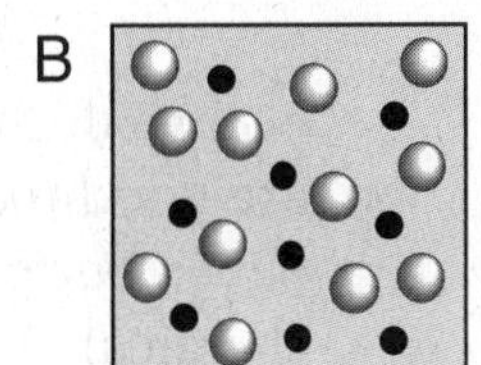

C

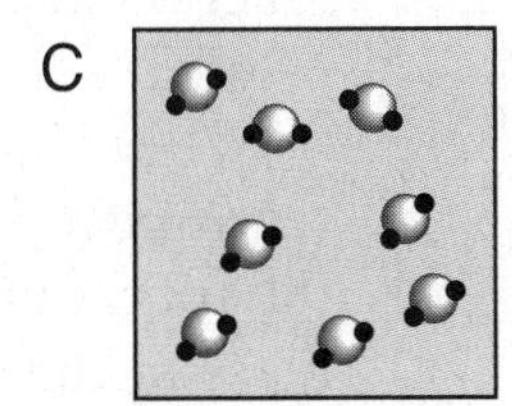

D

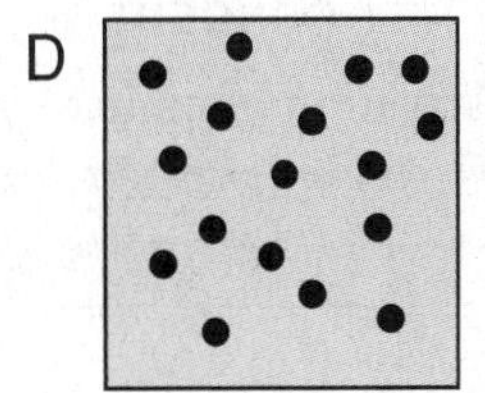

See page 132 for the answer.

PHYSICAL AND CHEMICAL CHANGES

The matter all around us is always changing. There are two basic types of changes: physical and chemical changes. When a **physical change** occurs, the composition of the substance stays the same, but the shape, phase, or location may be altered. When water changes to ice or to water vapor, a physical change has taken place. Although the water is in a different state of matter, it is still water and still has the same chemical composition. If a piece of paper is torn into many tiny pieces, a physical change has taken place. Although the size and shape of the paper has changed, it is still paper. No chemical reactions have occurred. A physical change happens when mixtures are created or separated. Also, if a substance is moved to another location or if there is a change in shape, then a physical change has occurred. For example, gold undergoes a physical change when it is hammered into a ring or another piece of jewelry.

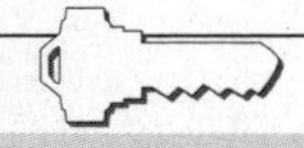

KEY CONCEPT

During a physical change, the composition of the substance stays the same, but the shape, phase, or location may be altered.

A **chemical change** or reaction occurs when two or more materials react and a new substance that has different properties than the original materials forms. Additionally, energy is often absorbed or released in a chemical reaction. It is usually easy to recognize if a chemical change has occurred. Bubbles, the release of heat or smoke, or a color change are good indicators or evidence of a chemical reaction. A piece of paper burning, a nail rusting, and a cake baking are all examples of chemical changes.

The **law of conservation of matter** states that matter cannot be created or destroyed. Because matter cannot be created or destroyed, the products of a chemical reaction always have the same mass as the original substances.

KEY CONCEPT

The law of conservation of matter states that matter cannot be created or destroyed by ordinary means.

When materials interact, atoms are rearranged into new compounds; however, the total number of atoms and the total mass of the system remains the same as the original system. This can be illustrated using a chemical equation. A **chemical equation** is a representation or a model of a chemical reaction. Atomic symbols and chemical formulas are used in chemical equations. Let's look at an example. Hydrogen and oxygen gas chemically react to form water. The chemical equation for this chemical reaction is

$$2H_2 + O_2 \rightarrow 2H_2O$$

The hydrogen gas (H_2) and the oxygen gas (O_2) are called the **reactants**, and the water (H_2O) is the **product** in this chemical reaction. The number 2 that is found before the chemical formula for hydrogen gas is called a **coefficient**. This indicates that when two hydrogen gas molecules react with one oxygen gas molecule, two molecules of water are created. The total number of atoms and their mass after the reaction is the same as the mass and number of atoms before the reaction. In this chemical equation, there are six atoms on the reactant side of the equation. On the product side of the equation, there are two molecules of water; therefore there are four atoms of hydrogen and two atoms of oxygen. The law of conservation of matter requires that every atom that appears on the reactant side of a chemical reaction must also appear on the product side; only their combination and arrangement can change.

ENERGY

After reading this section you should be able to:

- Describe the various forms of energy and trace energy transformations from one form to another.
- Describe how heat is transferred from one object to another through conduction, convection, and radiation.
- Recognize that the Sun is a major source of energy and that its energy consists of varying wavelengths of electromagnetic radiation.
- Demonstrate how light interacts with matter through transmission, scattering, and absorption

FORMS OF ENERGY

Energy causes things to happen in our world. The Sun gives us light and heat energy. Energy powers our televisions and computers. The food we consume provides us with energy to work and play. We use the term "energy" often, but scientifically speaking, energy has a specific meaning; **energy** is the ability to do work. In order to do work, a force must be exerted over a distance. When you lift a book, run around, and even move your pencil as you write, you are doing work, and hence using energy.

> **KEY CONCEPT**
> Energy is the ability to do work.

Energy is a property of many substances, and it takes many forms. Heat, light, electricity, chemical, nuclear, sound, and mechanical motion are all forms of energy. Let's examine these forms of energy a little closer. All matter has heat energy, or thermal energy. Matter is composed of tiny atoms and molecules that are always in motion. The more heat energy a sample of matter has, the faster the atoms and molecules move. Light and electricity are forms of electromagnetic energy. Light consists not only of the visible light and colors we see but also other forms of electromagnetic energy, like ultraviolet light, infrared light, and microwaves that travel through the universe. Electricity is the movement of electrons through a conductor, like copper wires. Light bulbs, computers, and televisions all use electrical energy to do work. Chemical energy is the energy that is stored in the bonds between atoms of compounds. During chemical reactions, these bonds can be broken, and the energy can be released. For example, when we digest our food, the chemical compounds, like sugars and carbohydrates, in the food undergo a chemical reaction. Some of the chemical bonds are broken, and the energy is released. This is how we get energy from our food. Also, chemicals like gasoline also store energy in chemical bonds. When gasoline is burned, a chemical reaction takes place, and the energy that is stored within it is released. Energy is found not only in chemical compounds but also in the nucleus of atoms. Nuclear energy is stored in the nucleus of atoms. This energy can be released by splitting atoms in a nuclear power plant. Mechanical energy is the energy of moving objects. When an object is moving, it has mechanical energy. A hammer being swung, a gymnast doing flips, and a paper plane flying through the air all have mechanical energy. Sound is a special type of mechanical energy. When an object vibrates, sound waves that can move, or travel through matter, are generated. Our ears can sense these sound waves.

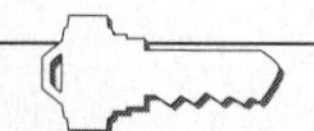

KEY CONCEPT

Energy is a property of many substances, and it takes many forms. Heat, light, electricity, chemical, nuclear, sound, and mechanical motion are all forms of energy.

ENERGY TRANSFORMATIONS

The energy in our universe exists in many forms as just described. This energy doesn't run out. Similar to the law of conservation of matter, the law of conservation of energy states that energy cannot be created or destroyed by ordinary means, but that it can be changed from one form to another. Energy transformations are always occurring around us and even within our bodies. For example, the chemical energy stored within the food we eat is transformed to mechanical energy when we use it to move and play.

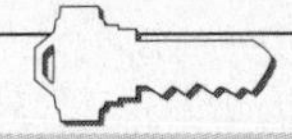

KEY CONCEPT

The law of conservation of energy states that energy cannot be created or destroyed, but that it can be changed from one form to another.

The electrical energy that runs through power lines is transformed into light and heat energy when we plug in a lamp. The chemical energy in batteries can be used in a flashlight, which converts it to light and heat energy.

Example 8

Which forms of energy does a burning candle release?

A. electricity and chemical

B. chemical and mechanical

C. light and heat

D. sound and electricity

See page 132 for the answer.

Because of our reliance on electricity, we are dependent on many energy sources, like fossil fuels, to harness energy and convert it to electricity. Fossil fuels, like coal, oil, and natural gas, were formed millions of years ago from living matter and store chemical energy. In many power plants, fossils fuels are burned. The chemical energy in the fossil fuel is transformed into heat energy. This heat energy is used to boil water. The steam rising from the boiling water turns a fan-like piece of equipment called a turbine; the energy has been converted to mechanical energy. The spinning turbine is connected to an electric generator. A generator consists of a long coiled wire surrounded by a magnet. As the generator turns, an electric current is produced in the wires, converting the mechanical energy to electricity. The electricity then travels through power wires to homes and buildings to provide us with energy for running our appliances. (See the figure on page 114 of a coal-burning power plant.)

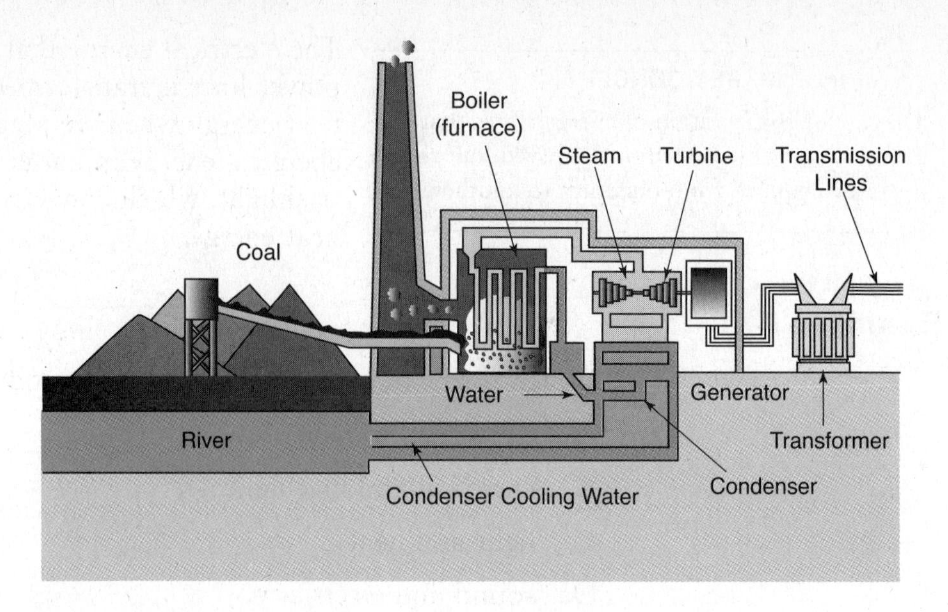

Scientists are starting to use other energy sources to generate electricity because the burning of fossil fuels can create harmful pollution and because they are nonrenewable. Other energy sources, like wind and moving water, can be used to spin turbines and generators. Additionally, nuclear energy is used in some power plants to generate electricity. Fission, or the splitting of atoms, releases a tremendous amount of energy from the nucleus of atoms. The energy is used to boil water and spin a turbine and generator. In nuclear power plants, nuclear energy is transformed to heat, then mechanical, and finally electrical energy.

THE TRANSFER OF HEAT ENERGY

Heat energy moves in predicable ways. It flows through materials and even across space from warmer objects to cooler ones. There are three ways that heat is transferred: conduction, convection, and radiation. Conduction is the direct transfer of heat from one material to another. If a surface is hot and you touch that surface with your hand, then your hand will get warm. This occurs as a result of the collisions between atoms and molecules. Convection is the transfer of heat within a fluid. This movement of heat occurs in fluids like the ocean and even in the mantle of the Earth and in the atmosphere, which both behave like fluids. Warm, less dense material rises, and cooler, denser material takes its place. As it warms, it becomes less dense and rises. This is called convection currents and results from the difference in temperature within fluid materials. Radiation is the transfer of heat through empty space.

The Sun's heat energy travels through space and to the Earth by radiation. This radiation can be carried by waves or particles.

In the figure below, wood is being burned. If you hold your hands over the fire, you will feel heat because of convection. The air above the fire is being warmed. This warm air rises because it is less dense. If you place a metal poker in the fire, it will get hot because of conduction. A glove should be worn to protect your hand from heat conduction as you hold the hot metal poker. Finally, if you place your hands beside the fire, you will feel the heat by radiation. The heat travels through empty spaces between the gas molecules in the air in all directions and toward your hands.

> **KEY CONCEPT**
>
> Heat can be transferred from one object to another through conduction, convection, and radiation.

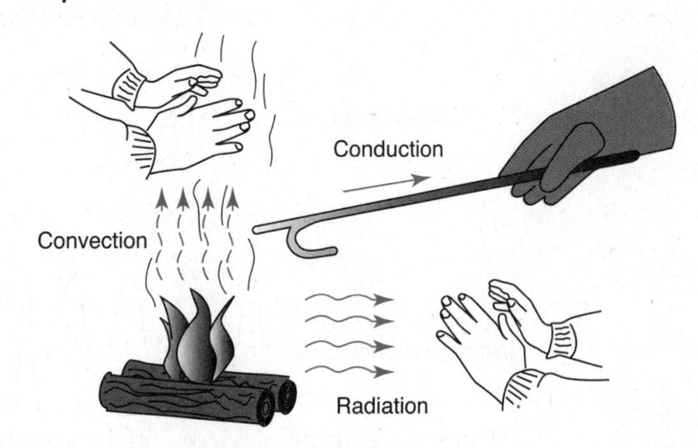

Example 9

How is solar energy from the Sun transferred to the Earth?

A. conduction

B. convection

C. radiation

D. refraction

See page 132 for the answer.

MOVEMENT OF ENERGY IN WAVES

Some forms of energy, like electromagnetic radiation and sound, travel as waves. Waves are traveling disturbances that carry energy. Energy and vibrations in materials can generate waves that spread away from the source and that transfer energy from one place to another. Ocean waves, seismic waves, and sound are examples of **mechanical waves**. Mechanical waves require a **medium**, or substance, to travel through. For example, ocean waves travel through a medium of water, and seismic waves from earthquakes travel through a medium of rock and soil. Waves move at different speeds in different materials.

Mechanical waves are classified based on how they move and transfer energy. Two of the most common classifications of mechanical waves are transverse and longitudinal. In a **transverse wave** the matter in the medium moves up and down, or perpendicular to the direction of the wave. In **longitudinal**, or **compression waves**, the matter in the medium moves back and forth in the same direction of the wave. (See the figure.)

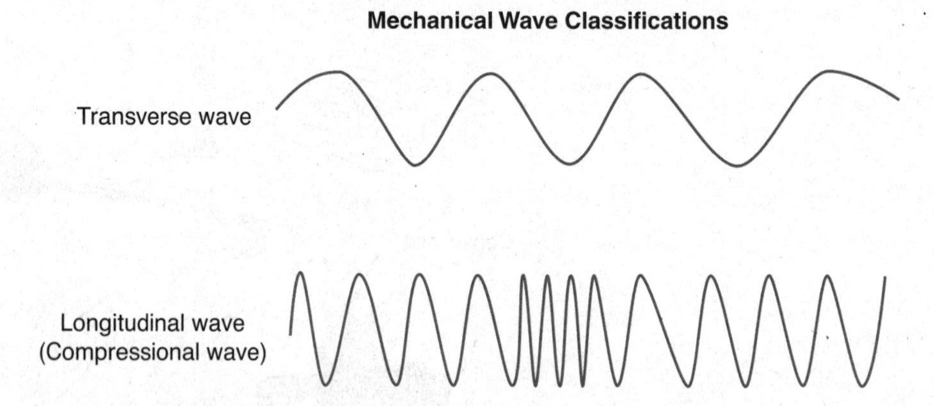

When describing a wave, scientists often use the terms wavelength and amplitude. The topmost part of a wave is called the **crest** and the bottommost part of the wave is the **trough**. The **wavelength** is the distance between any two adjacent corresponding locations of a wave, for example, crest to next crest or trough to next trough. The **amplitude** is the height of the wave from the starting undisturbed position. (See the figures.)

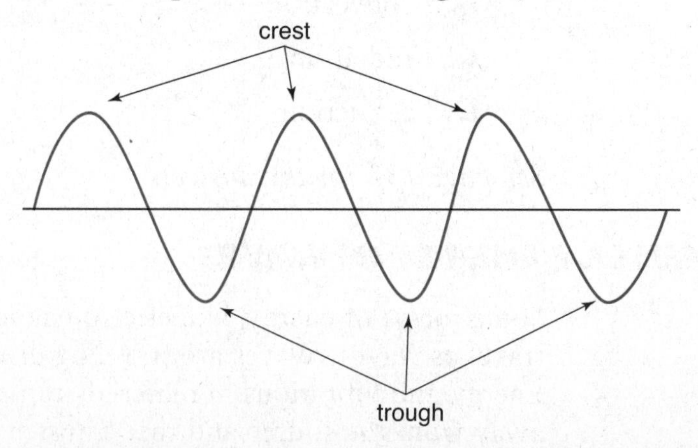

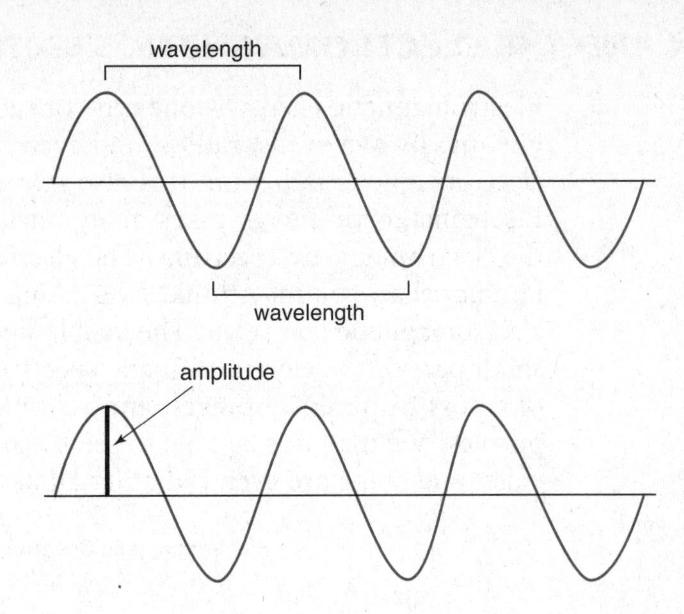

Example 10

Which of these diagrams shows the wave with the longest wavelength?

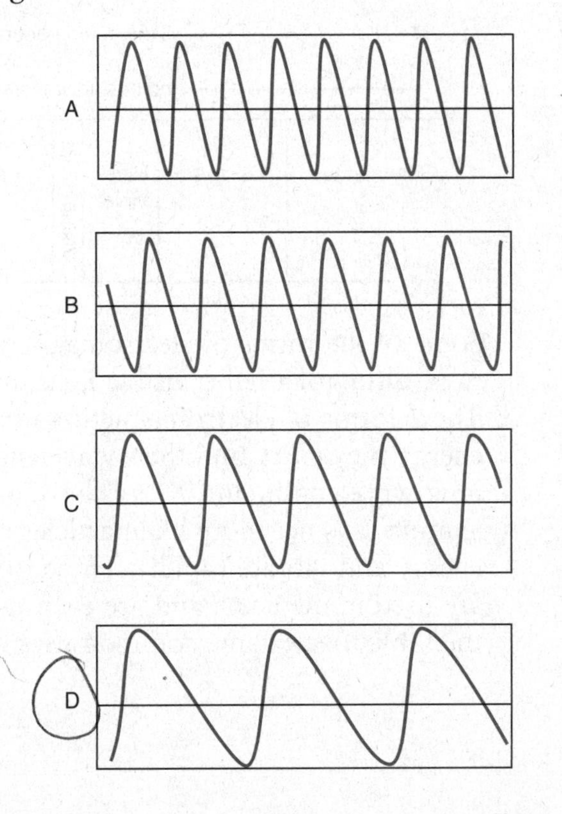

See page 133 for the answer.

THE SUN AND THE ELECTROMAGNETIC SPECTRUM

Electromagnetic energy is one type of radiation. Many appliances, like microwave ovens, radios, and even cell phones, use electromagnetic radiation. It is also released by the Sun. Electromagnetic energy takes many forms and can be illustrated by the electromagnetic spectrum. The electromagnetic spectrum is more familiar than you may think. Everything that you see is due to the electromagnetic spectrum. The visible light that our eyes sense is a small part of the electromagnetic spectrum; it consists of the rainbow of colors from reds, oranges, and yellows to greens, blues, and purples. We may not be able to see the other forms of light, but they exist, and some are even radiating all around us. Examine the figure.

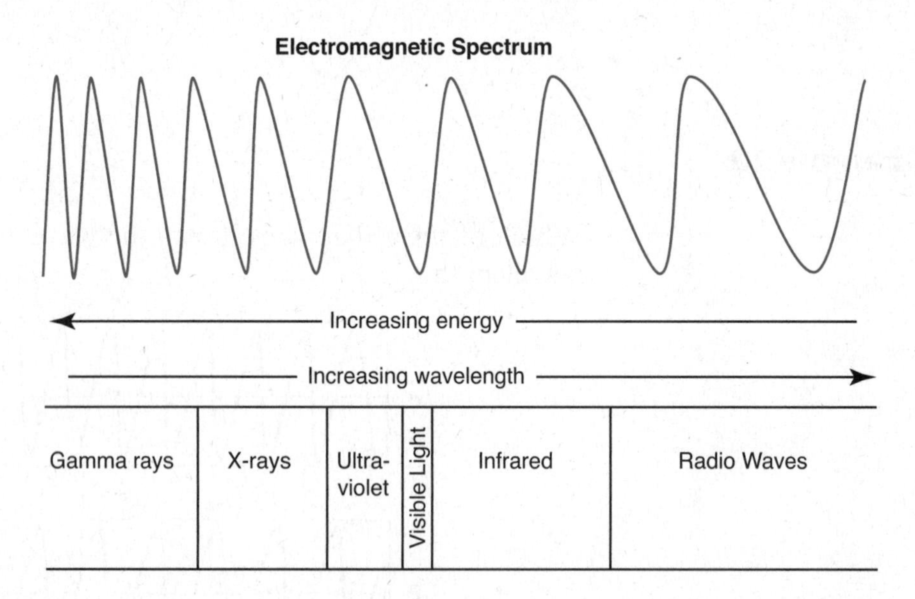

Some of the forms of electromagnetic radiation are gamma rays, X-rays, ultraviolet light, visible light, infrared light, and radio waves. These forms of electromagnetic radiation differ due to the amount of energy they carry and their wavelengths. Radio waves have a very long wavelength, and X-rays have a shorter wavelength. Actually, gamma rays act more like particles than waves; they carry lots of energy and vibrate rapidly. These forms of electromagnetic energy are used in medicine and are even used to carry information. Look at the table to see some common uses of electromagnetic radiation.

COMMON USES FOR ELECTROMAGNETIC RADIATION

Type of Electromagnetic Radiation	Common Uses
Gamma rays	Used in medicine to kill or treat cancers and tumors. Used to kill pests and insects in food.
X-rays	Used in medicine because X-rays can pass through skin, fat, and other soft tissue, but not through bones. They are also used for security purposes to scan suitcases and bags to reveal any dangerous contents.
Ultraviolet Light	UV light causes skin cancer. Most UV light from the sun is absorbed by the ozone (O_3) layer.
Visible Light	We can see the colors of the visible spectrum with our eyes: red, orange, yellow, blue, indigo, and violet.
Infrared Light	We feel this heat. Warm objects radiate infrared light. It can be sensed using infrared, or night vision, goggles.
Radio Waves	AM/FM broadcast radios, televisions, cell phones, baby monitors, microwave ovens.

The Sun is a major source of the Earth's energy, and solar energy consists of electromagnetic energy with a range of wavelengths. Only a tiny fraction of the energy emitted by the Sun reaches the Earth. The Sun's energy arrives as light, transferring energy from the Sun to the Earth with a variety of wavelengths, consisting of visible light, infrared light, and ultraviolet radiation.

KEY CONCEPT

The Sun is a major source of the Earth's energy. Solar energy consists of electromagnetic energy with a range of wavelengths, including visible light, ultraviolet light, infrared, radio waves, and gamma rays.

Example 11

Using the diagram of the Electromagnetic Spectrum in the figure on page 118, which form of electromagnetic radiation has the longest wavelength?

A. radio waves

B. visible light

C. gamma rays

D. ultraviolet light

See page 133 for the answer.

LIGHT ENERGY

The Sun and other sources like light bulbs and television screens emit light. Other objects do not emit their own light, yet we can still see them. To see an object, light from that object, either emitted by it or scattered off it, must enter our eyes. Light interacts with matter in a variety of ways. Matter can transmit, absorb, or scatter light. Some matter allows light to **transmit**, or pass through it. Water and glass transmit light. (See the figure.)

Transmission of Light

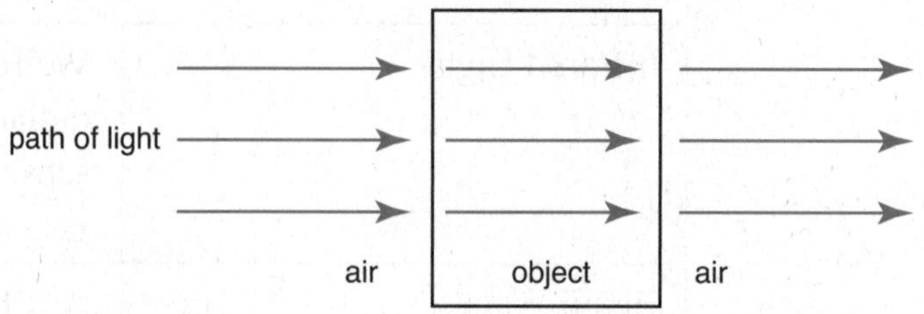

path of light
air
object
air

Sometimes when light transmits through an object, the rays of light will bend; this is called refraction. **Refraction** occurs because light moves faster in some materials. When light passes through a medium at an angle and comes into contact with a different medium it refracts, causing the light to bend in a different direction. (See the figure.)

Refraction of Light

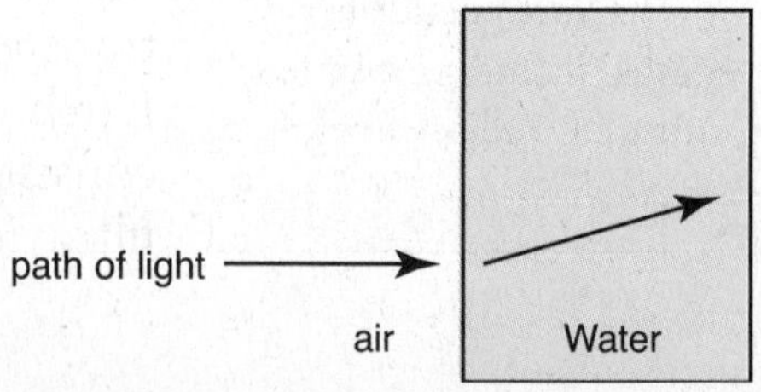

path of light
air
Water

Some materials absorb light. Absorption occurs when light strikes a medium and is taken into that medium. On a sunny day, you have probably observed, that if you are wearing a dark-colored shirt you will feel warmer. This is because the dark shirt absorbed the light, and the energy was transformed to heat. (See the figure.)

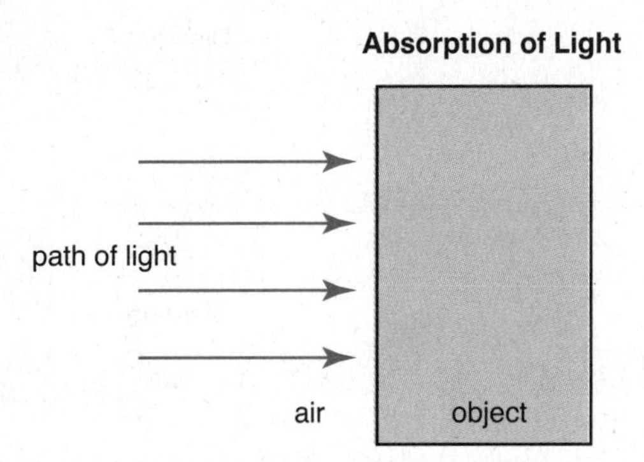

When light bounces off an object, it is called scattering. Some objects scatter light randomly in many directions. Other objects with smooth surfaces scatter the light at the same angle at which it reached its surface; this is called reflection. Reflection occurs when light strikes a surface at an angle and then bounces off at an equal angle. (See the figure.)

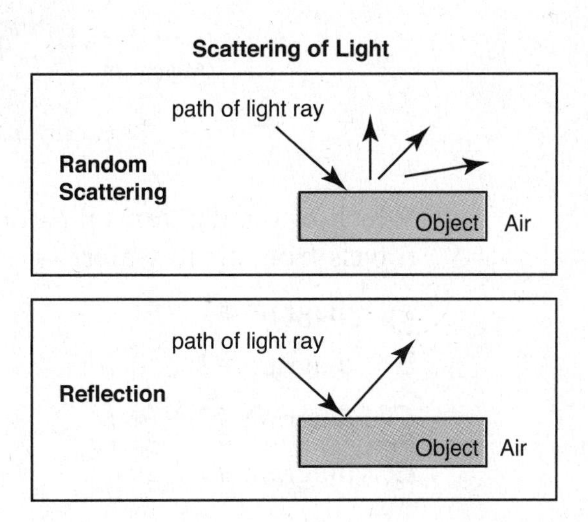

Example 12

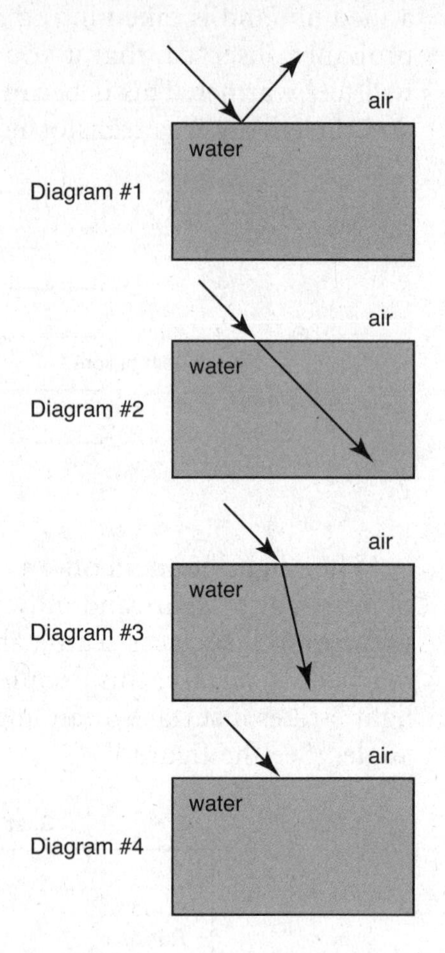

Which of the diagrams illustrates the refraction of light as it travels from air to water?

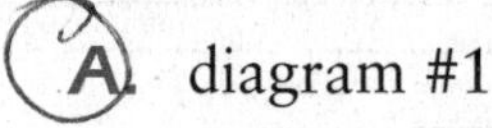

A. diagram #1

B. diagram #2

C. diagram #3

D. diagram #4

See page 133 for the answer.

Objects are classified based on how they interact with light. They can be classified as transparent, translucent, or opaque. Transparent objects allow most of the light that interacts with it to transmit through them. Clear glass and water can be transparent. Translucent objects allow some light to transmit and some light to scatter. Waxed paper and frosted glass are translucent. Finally,

opaque objects, like books, tables, and walls, scatter and/or absorb all of the light that strikes them.

Colors may appear as a result of light interactions. The light from the Sun appears white but is composed of all of the colors of the visible spectrum (red, orange, yellow, green, blue, indigo, and violet). This can be observed by placing a prism, or a piece of transparent glass, in the beam of sunlight. The colors of the spectrum vary in wavelength; red has a longer wavelength, while violet has a shorter wavelength. When light strikes an object, some wavelengths of light may be absorbed, while others are scattered. The color that we perceive an object to be is actually the color that the object scatters. For example, a red apple absorbs all of the colors of the spectrum and scatters the color red. That red light enters our eyes, and we perceive the apple as red. (See the figure.)

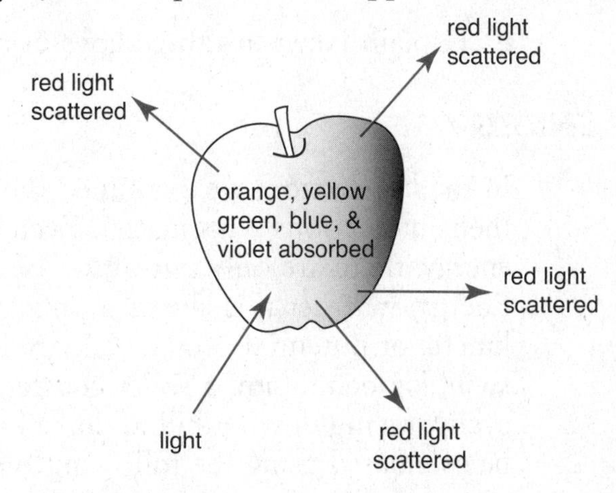

A black object absorbs all of the colors of the spectrum, while a white object reflects all of the colors. Translucent objects appear to be the color that is transmitted through them. For example, a translucent piece of green glass absorbs all of the colors of the spectrum but allows green light to transmit through it. (See the figure.)

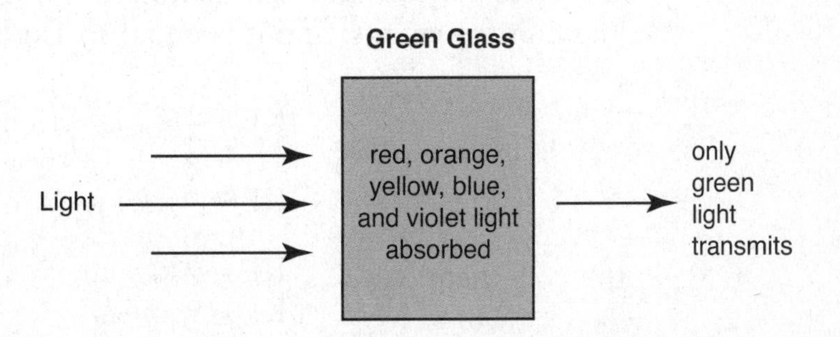

MOTION AND FORCES

After reading this section you should be able to:

- Differentiate between the two states of energy, potential and kinetic.
- Recognize that motion can be hindered by forces such as friction and air resistance.
- Recognize that all objects exert a gravitation force on other objects and that the magnitude of the force is dependent on the mass of the object and the distance between them.
- Understand the effects of balanced and unbalanced forces on the speed and direction of objects.
- Explain Newton's three laws of motion.

STATES OF ENERGY

In the last section, we examined the various forms of energy and their interactions with matter. Even though there are many forms of energy, there are only two states of energy, kinetic and potential. Energy, whether it is chemical, nuclear, or mechanical, is either kinetic or potential. Kinetic energy is the energy of motion. A swinging pendulum, a roller coaster speeding on its track, and a cyclist racing down a hill all have kinetic energy. Kinetic energy can be calculated using the following formula:

$$\text{Kinetic energy} = \frac{1}{2} \times \text{Mass} \times \text{Velocity}^2$$
$$\text{KE} = \frac{1}{2}\text{mv}^2$$

If a ball with a mass of 2 kilograms is moving with a velocity of 3 meters per second, then its kinetic energy is 18 joules. A joule is the unit for energy, and it is equal to 1kg × m^2/s^2.

$$\text{KE} = \frac{1}{2} \times 2 \text{ kg} \times (3 \text{ m/s})^2$$
$$\text{KE} = \frac{1}{2} \times 2 \text{ kg} \times 9 \text{ m}^2/\text{s}^2$$
$$\text{KE} = 9 \text{ kg} \times \text{m}^2/\text{s}^2$$
$$\text{KE} = 9 \text{ joules}$$

Kinetic energy is dependent on both mass and velocity. Velocity, though, has more of an affect on the kinetic energy of an object.

Potential energy is stored energy. Food, gasoline, and a compressed spring all have potential energy. One type of potential energy is called **gravitational potential energy**. Gravitational potential energy is stored energy due to its position, or height. A diver standing on a high diving board has gravitational potential energy. The higher an object is off the ground, the more gravitational potential energy the object has. Gravitational potential energy can be calculated using the following formula:

$$\text{Gravitational potential energy} = \text{Mass} \times \text{Gravity} \times \text{Height}$$

Gravity in this equation refers to the acceleration due to gravity or 9.8 m/s^2. Let's look at an example. If a diver with a mass of 65 kilograms is standing on a diving board 10 meters above the ground, his gravitational potential energy would be

$$\text{GPE} = 65\text{ kg} \times 9.8\text{ m/s}^2 \times 10\text{ m}$$
$$\text{GPE} = 6370\text{ J}$$

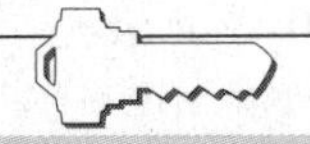

KEY CONCEPT

Energy is either kinetic or potential. Kinetic energy is the energy of motion. Potential energy is stored energy, or energy of position.

Just at energy is transformed from one form to another, energy also is transformed between states. For example, when a rubber band is stretched, its potential energy increases. As the rubber band is released, the potential energy is transferred to kinetic energy. Energy transformations also occur as a roller coaster travels along its track. Look at the figure on page 126. The roller coaster begins its trip at point A. As it travels up the hill to point B its potential energy increases because its height is increasing. As it travels down the first hill, the potential energy is converted to kinetic energy. At point C, the roller coaster is moving the fastest and has a lot of kinetic energy. Its kinetic energy drives the coaster up the next hill to point D. The roller coaster slows down as it goes up the hill; it loses kinetic energy and gains potential energy. Along the entire roller coaster ride, its greatest potential energy is at point B because it is at the highest point; its greatest kinetic energy is at point C because it has the highest velocity.

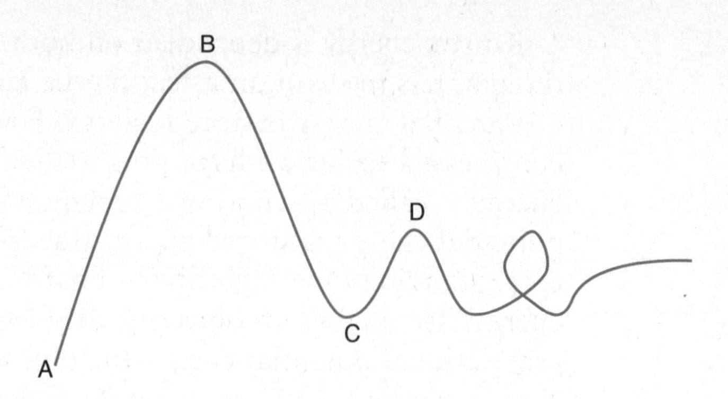

FORCES AND FRICTION

Forces act on objects to make them move. A **force** is a push or a pull that can cause an object to accelerate. If a child pushes a toy car, then the car may move. Forces, such as friction and air resistance, can also hinder, or slow down, the motion of objects. **Friction** acts in the direction opposite to the motion of the object; it is the force that resists motion when the surface of one object comes into contact with the surface of another object. Friction is due to the attractive forces between the surfaces. Because of friction, we are able to walk around. Without friction between the soles of our shoes and the ground, we would slip and fall. If a force is applied to a wooden block, the force friction will act in the opposite direction. The force of friction will eventually cause the wooden block to stop moving. (See the figure.)

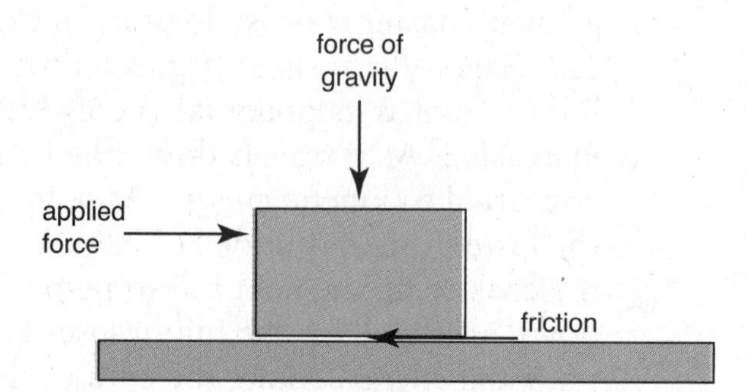

When more than one force acts on an object at the same time, the forces can either reinforce or cancel each other. If two equal forces act on an object in opposite directions, they will cancel each other out, and the object will not move; the forces are balanced. If two unequal forces act on an object in opposite directions, the unbalanced forces will change the speed

> **KEY CONCEPT**
>
> A force is a push or a pull that can cause an object to accelerate. Motion can also be hindered by forces, such as friction and air resistance.

and/or direction of the object. If more than one force acts on an object from the same direction, the forces will be unbalanced and reinforce one another; this will cause the object to change its speed and/or direction. Balanced forces will not cause a change in the speed or direction of an object's motion. Unbalanced forces will cause changes in the speed or direction of an object's motion. (See the figure.)

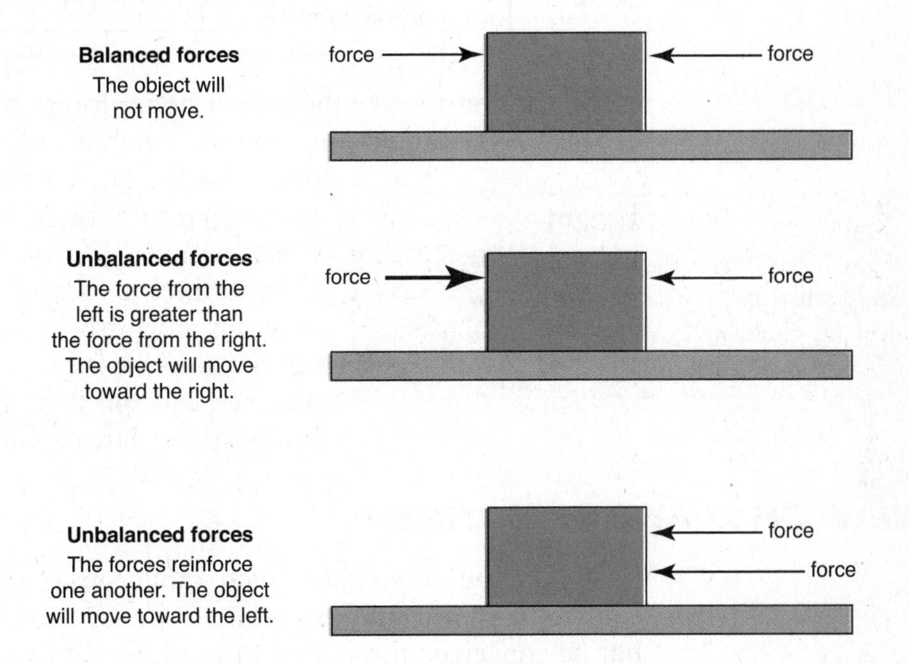

GRAVITY

Gravity is another force that affects all objects. Every object exerts a gravitational force on every other object. The force of gravity depends on the mass of the objects and the distance between them. As the mass of one or both objects increases, so will the gravitational force that exists between them. Alternatively, the gravitational force will decrease as the distance between the objects increases. (See the figures.)

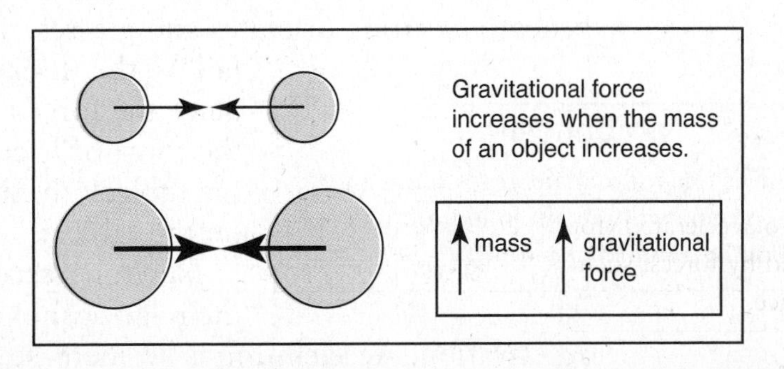

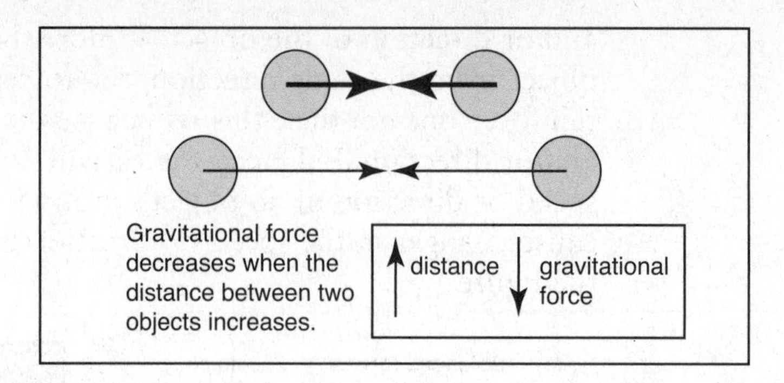

We cannot observe the gravitational forces between most objects because everything on or near the Earth is pulled toward the center of the Earth. The Earth has a tremendous mass; therefore, the gravitational force it exerts on all objects is so great that we cannot observe the gravitational forces between most other objects. This relationship between gravity, mass, and distance was put forth by Isaac Newton and is named the **law of universal gravitation.**

> **KEY CONCEPT**
>
> Every object exerts a gravitational force on every other object. The force of gravity depends on the mass of the objects and the distance between them.

NEWTON'S LAWS OF MOTION

Isaac Newton's discoveries were numerous. Not only did he contribute to our understanding of gravity, mathematics, and light, but he conceived three **laws of motion** that explain the motion of objects in the universe. The first law of motion states that an object at rest will remain at rest and an object moving in a straight line at a steady speed will continue to move in a straight line at a steady speed unless an unbalanced force acts on it. An object that is not being subjected to a force will continue to move at a constant speed and in a straight line. This law involves the concept of inertia. **Inertia** is the tendency of an object to resist change in motion. If an object is at rest, it will resist movement. And, if an object is in motion, it will resist any change in acceleration or deceleration. Perhaps you have seen a performer swiftly pull a tablecloth from beneath a setting of plates and glasses. The plates and glasses do not fall on the floor because of their inertia. Since the force of the inertia is greater than the force of friction between the plates and the tablecloth, the plates and glasses remain on the table.

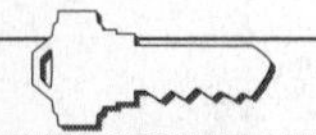

> **KEY CONCEPT**
>
> An object at rest will remain at rest, and an object in motion will remain in motion unless acted upon by an unbalanced force.

Newton's second law of motion involves the relationship between force, mass, and acceleration. **Acceleration** is an increase in the rate or speed of an object. Newton's second law can be expressed using the following formula:

$$\text{Force} = \text{Mass} \times \text{Acceleration}$$

In plain terms, this law explains that it takes a greater force to accelerate an object with a larger mass than one with a smaller mass. If you were to drop a ping pong ball and a bowling ball from the same height at the same moment, which would hit the ground first? Actually, they would hit the ground at the same moment because gravity accelerates all objects at the same rate. But, they would hit the ground with different forces. Because the bowling ball has a larger mass than the ping pong ball, it will hit the ground with a greater force when accelerated due to gravity. This is an example of Newton's second law of motion.

Newton's third law of motion states that for every action there is an equal but opposite reaction. If you apply a force to a wall by pushing against it, the wall exerts an equal force back in the opposite direction. All forces act in pairs, and these forces are equal in magnitude but opposite in direction. Imagine a rocket blasting off the ground. The hot gases and flames shoot downward, while the rocket is thrust upward. Or consider a balloon filled with air. If the balloon is released, that air will rush out of it in one direction, and the balloon will move in the opposite direction. These are both examples of Newton's third law of motion.

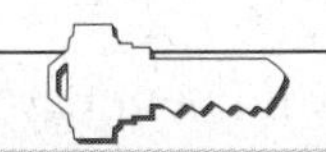

KEY CONCEPT

All forces act in pairs, and these forces are equal in magnitude but opposite in direction.

SUMMARY

In this chapter, we explored the basic principles of matter, including interactions between energy and matter, the physical principles of motion that operate in our universe, and the laws that govern how and why objects move.

Everything around us is matter. Matter is anything that takes up space and has mass. There are three basic phases of matter: solid, liquid, and gas. Matter is always changing. There are two basic types of changes, physical and chemical changes. Physical properties of matter can be measured and observed; they can include the boiling point, melting point, and density.

Every substance we observe is either an element or a compound. Elements are simple substances. They have been arranged in a table called the Periodic Table of Elements by their atomic numbers.

Theses elements have been classified into three categories based on their similar properties: metals, nonmetals, and noble gases.

Atoms are the fundamental particles of an element and consist of protons, neutrons, and electrons. Atoms join together to form molecules of various compounds. A molecule is composed of two or more atoms.

A compound is a substance made of two or more elements. Most of the materials around us are compounds. Many of the substances we encounter are mixtures of elements and compounds. A mixture is a combination of two or more substances in varying amounts. A mixture can usually be separated into its original substances using one or more of their characteristic properties.

Energy is the ability to do work and is a property of many substances. Heat, light, electricity, chemical, nuclear, sound, and mechanical motion are all forms of energy. The law of conservation of energy states that energy cannot be created or destroyed by ordinary means, but that it can be changed from one form to another. Energy transformations are always occurring around us and even within our bodies. Heat energy moves in predictable ways. There are three ways that heat is transferred: conduction, convection, and radiation. Other forms of energy, like sound, travel as waves, or traveling disturbances.

The Sun is a major source of the Earth's energy, and solar energy consists of electromagnetic energy with a range of wavelengths. Only a tiny fraction of the energy emitted by the Sun reaches the Earth, transferring energy from the Sun to the Earth. The Sun's energy arrives as light with a variety of wavelengths, consisting of visible light, infrared light, and ultraviolet radiation.

To see an object, light from that object, either emitted by it or scattered off of it, must enter our eyes. Light interacts with matter in a variety of ways. Matter can transmit, absorb, or scatter light. Objects are classified based on how they interact with light. They can be classified as transparent, translucent, or opaque. Colors may appear as a result of these light interactions.

There are two states of energy, kinetic and potential. Kinetic energy is the energy of motion. Potential energy is stored energy. Gravitational potential energy is stored energy due to its position, or height. Energy can be transformed from one state to another.

A force is a push or a pull that can cause on object to move. Motion can also be hindered by forces, such as friction and air resistance. Friction acts in the direction opposite to the motion of the object. Balanced forces will not cause a change in the speed or direction of an object's motion. Unbalanced forces will cause changes in the speed or direction of an object's motion.

Every object exerts a gravitational force on every other object. The force of gravity depends on the mass of the objects and the distance between them. As the mass of one or both objects increases, so will the gravitational force that exists between them.

Isaac Newton conceived three laws of motion that explain the motion of objects. The first law of motion states that an object at rest will remain at rest and an object in motion will remain in motion unless an unbalanced force acts upon it. The second law of motion involves the relationship between force, mass, and acceleration. The third law of motion states that for every action there is an equal but opposite reaction.

KEY TERMS FROM THIS CHAPTER

matter
physical properties
boiling point
melting point
density
mass
volume
conductivity
element
compound
atom
nucleus
protons
neutrons
electrons
Periodic Table of Elements
groups
periods
atomic number
chemical symbol
atomic weight
metal
luster
malleable
ductile
non-metal
noble gases
molecules
chemical formula
solid
liquid
gas
mixture
solution
soluble
solute
solvent
physical change
chemical change
law of conservation of matter
chemical equation
reactants
products
coefficient
energy
heat energy
electromagnetic energy
electricity
chemical energy
nuclear energy
mechanical energy
sound
law of conservation of energy
generator
fission
conduction

convection
convection currents
radiation
waves
mechanical waves
medium
transverse wave
longitudinal wave
compression wave
crest
trough
wavelength
amplitude
electromagnetic spectrum
transmission
refraction
absorption
scattering
reflection
transparent
translucent
opaque
kinetic energy
potential energy
gravitational potential energy
force
friction
balanced forces
unbalanced forces
gravity
law of universal gravitation
Newton's laws of motion
inertia
acceleration

ANSWERS TO EXAMPLES

1. **D** To determine the density divide the mass (18 grams) by its volume (4 cm × 3 cm × 3 cm = 36 cm^3). D = 18 g/36 cm^3 = 0.5 g/cm^3.
2. **B** An atom is the smallest quantity of an element.
3. **D** Neon has the atomic number of 10, and therefore has ten protons in each atom.
4. **A** The substance exhibits the properties of a metal.
5. **B** H_2O, or water, is a compound because it is composed of more than one element (hydrogen and oxygen).
6. **C** As an ice cube melts, the molecules absorb heat energy and move farther apart as they change to the liquid phase of matter.
7. **B** Image B shows different particle types mixed together but not attached, or bonded. Images A and D only show particles of one type; this would represent an element. Image C represents particles of a compound.
8. **C** As a candle burns it releases energy in the form of light and heat.
9. **C** The energy from the Sun travels by radiation through empty space and to the Earth.

10. D The wave in diagram D has the longest wavelength, which is the distance between adjacent corresponding points on a wave.

11. A Radio waves have the longest wavelength, while gamma rays have the shortest wavelength.

12. C Diagram #3 shows the refraction or bending of light from air to water. In Diagram #1, the light bounces off or reflects off the surface of the water. In Diagram #2, the light transmits through the water at the same angle. The light stops and is perhaps absorbed by the water in Diagram #4.

PRACTICE QUIZ

For each of the questions or incomplete statements below, choose the best of the answer choices given. Turn to page 240 for the answers.

1. The data table displays the mass and volume of three different objects.

	Object A	Object B	Object C
Mass	12 g	6 g	5 g
Volume	6 cm^3	3 cm^3	5 cm^3

Using the formula:

$$\text{Density} = \frac{\text{Mass}}{\text{Volume}}$$

Which of the following statements is correct?

A. Object B is more dense than object A.

B. Objects A and B have equal densities.

C. Object C is more dense than object A.

D. Objects B and C have equal densities.

2. Atoms are composed of protons, neutrons, and electrons. What is the electric charge of the neutron?

A. positive

B. negative

C. neutral

D. double-charged

3. A battery is placed in a flashlight. The battery is attached to wires that connect to the light bulb. Which of the answers below shows the correct order of the energy changes?

A. chemical → electricity → light

B. chemical → mechanical → light

C. light → electrical → chemical

D. nuclear → mechanical → light

4. Using the diagram of the electromagnetic spectrum, which form of electromagnetic radiation has the shortest wavelength?

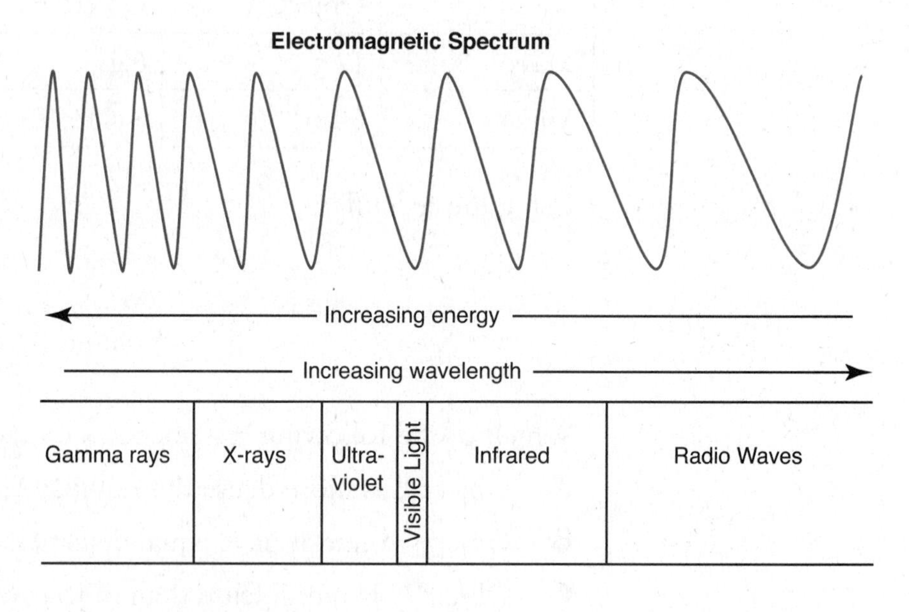

A. radio waves

B. visible light

C. gamma rays

D. ultraviolet light

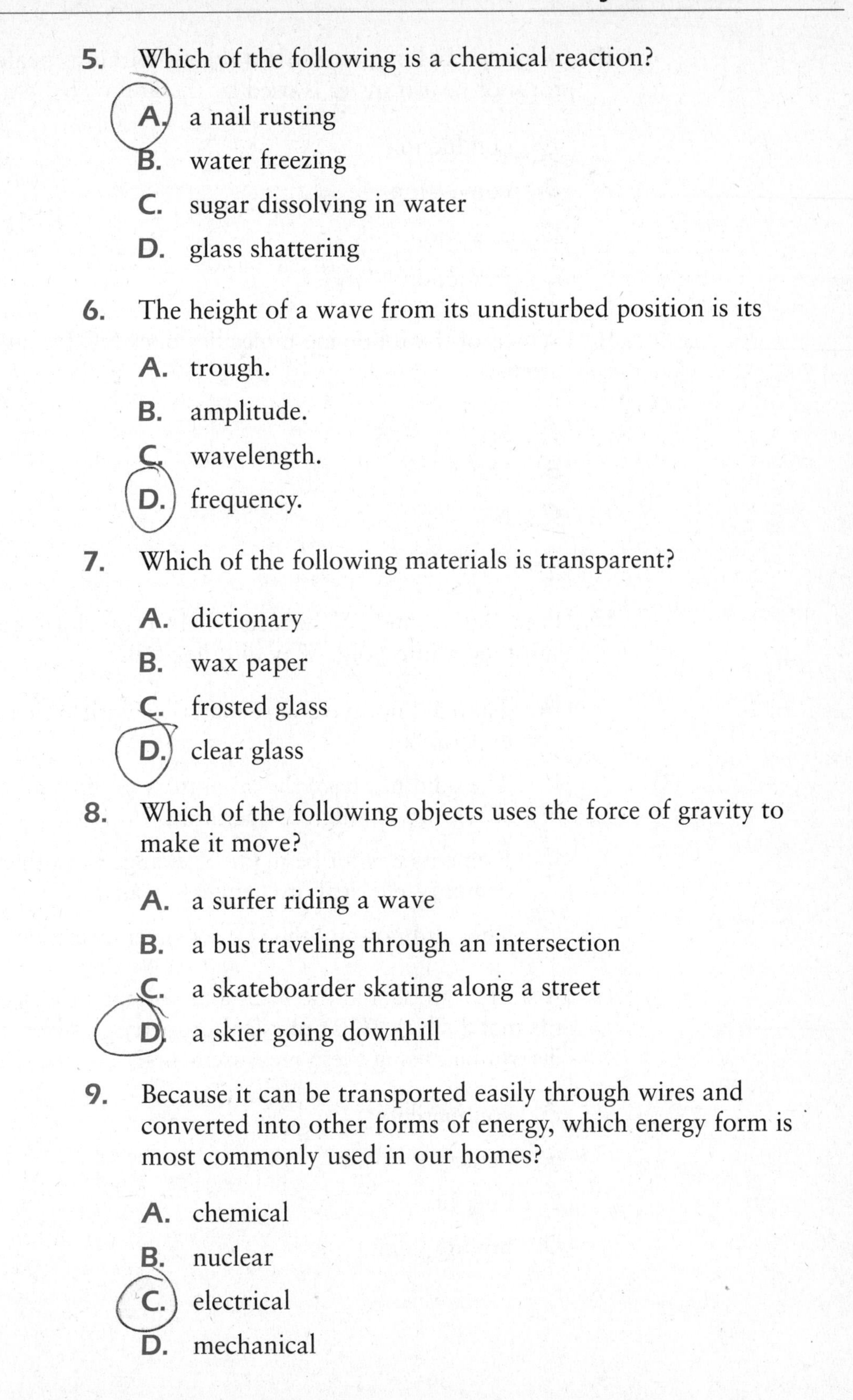

5. Which of the following is a chemical reaction?

   A. a nail rusting
   B. water freezing
   C. sugar dissolving in water
   D. glass shattering

6. The height of a wave from its undisturbed position is its

   A. trough.
   B. amplitude.
   C. wavelength.
   D. frequency.

7. Which of the following materials is transparent?

   A. dictionary
   B. wax paper
   C. frosted glass
   D. clear glass

8. Which of the following objects uses the force of gravity to make it move?

   A. a surfer riding a wave
   B. a bus traveling through an intersection
   C. a skateboarder skating along a street
   D. a skier going downhill

9. Because it can be transported easily through wires and converted into other forms of energy, which energy form is most commonly used in our homes?

   A. chemical
   B. nuclear
   C. electrical
   D. mechanical

10. A hot air balloon rises when the air within is heated. What form of heat transfer is used by the hot air balloon?

A. conduction

B. convection

C. radiation

D. refraction

11. Which of the following molecules does NOT contain two atoms?

A. O_2

B. NaCl

C. N_2

D. H_2O

12. In medieval times, alchemists tried to transform common substances into gold. Why did they fail?

A. They did not have pure materials with which to experiment.

B. They did not have the laboratory equipment to heat the substances to high temperature.

C. Elements cannot be made or changed to other elements by ordinary physical or chemical means.

D. They did not mix the correct substances to create gold.

13. A student measured the mass and volume of a cube made of the metal silver. What physical property of silver can be determined using these measurements?

A. conductivity

B. density

C. ductility

D. boiling point

14. When a liquid freezes, its molecules

  A. release heat energy and move farther apart.
  B. release heat energy and move closer together.
  C. absorb heat energy and move farther apart.
  D. absorb heat energy and move closer together.

15. Which of the diagrams shows the wave with the longest wavelength?

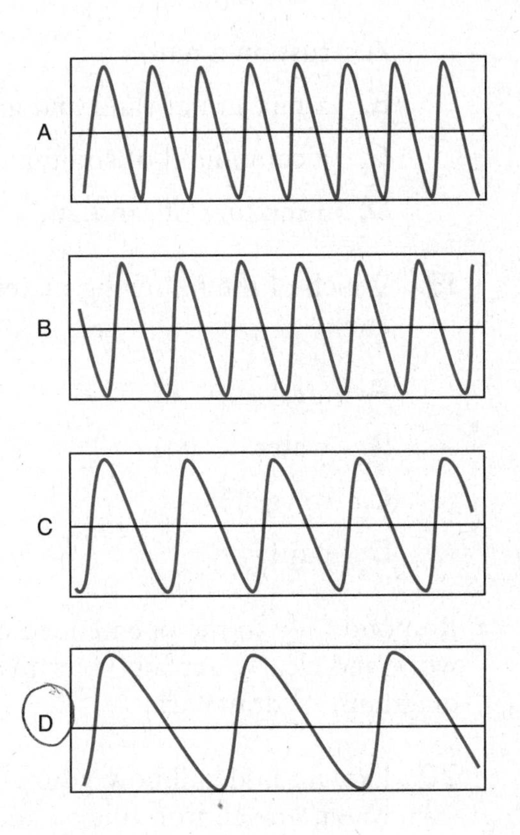

16. Which of the following is an example of potential energy being converted to kinetic energy?

  A. a roller coaster speeding down a tall hill
  B. a compressed spring
  C. a stretched rubber band
  D. a diver standing on a diving board

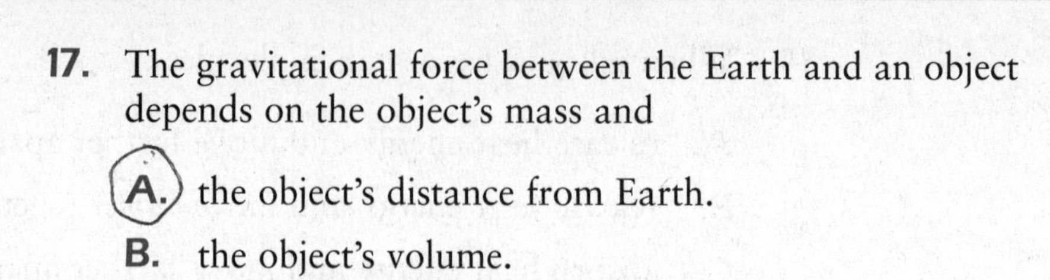

17. The gravitational force between the Earth and an object depends on the object's mass and

A. the object's distance from Earth.

B. the object's volume.

C. the object's state of matter.

D. the object's shape.

18. Which of the following can be separated by using a magnet?

A. rust on a nail

B. a mixture of sand and gravel

C. a compound of sodium and chlorine

D. a mixture of sand and iron filings

19. Which of the following materials is made of only one type of atom?

A. air

B. water

C. oxygen

D. paper

Respond fully to the open-ended question that follows. Show your work and clearly explain your answer. You may use words, tables, diagrams, or drawings.

20. Explain in detail how you would separate a mixture of sugar water, gravel, iron filings, and small wood particles. What properties of the substances are you using to separate the mixture? What supplies would you need and what steps would you take?

Chapter 5

EARTH AND SPACE SCIENCE

Our Earth is dynamic; it is always changing! Humans can observe some of these changes, while other changes happened thousands and even millions of years ago. Changes in the Earth system, like earthquakes, can occur rapidly, but some changes, like the rise and fall of sea level, happen very slowly. Fossils and layers of sedimentary rock provide evidence of these changes.

The Earth is the third planet from the Sun in our solar system, and the Earth is orbited by the Moon. The motion of the Sun, Earth, and Moon affects the Earth system; tides, seasons, and eclipses are caused by the motion in this Earth-Moon-Sun system. This chapter is divided into four sections:

- Weather and Climate
  - The Water Cycle
  - The Atmosphere
  - Global and Local Winds
  - Ocean Currents and Weather
  - Forecasting the Weather
  - Climate
- Soil, Minerals, and Rocks
  - Soil
  - Rocks and Minerals
  - Igneous Rocks
  - Sedimentary Rocks
  - Metamorphic Rocks
  - The Rock Cycle
- The Dynamic Earth
  - The Earth's Interior
  - Plate Tectonics
  - Earthquakes
  - Plate Boundaries
  - Weathering and Erosion

NOTE TO STUDENTS

Thoroughly read each section. As you read, pay close attention to the key concepts and try to answer the example questions found in each section. Answers to examples in this chapter are on page 181. At the end of the chapter, assess your knowledge of earth and space science with the practice quiz. Then, check your answers using the answer key provided on page 242.

- Using Fossils to Determine Earth History

- Earth in Space
  - Earth-Moon-Sun System
  - Phases of the Moon
  - Eclipses
  - Tides
  - Planets and Our Solar System
  - The Sun and Other Stars

WEATHER AND CLIMATE

After reading this section you should be able to:

- Describe and illustrate the water cycle.
- Recognize that air is a substance that surrounds us, takes up space, and circulates as wind.
- Understand that the uneven heating of the Earth causes areas of high and low pressure.
- Describe tools and techniques used to forecast weather.

THE WATER CYCLE

Most of the Earth's surface is covered by water. On Earth, water exists in all three basic states, or phases, of matter: solid, liquid, and gas. Water is observed as a solid in the form of ice and snow. Oceans, streams, and rain are composed of liquid water. In the atmosphere, water exists as a gas. Although the gas form of water is invisible, we can sense it on very humid days.

Water is found in Earth's oceans, rivers, lakes, glaciers, and even underground within soil and rocks. Water circulates through the Earth ecosystem. This circulation or movement of water through an ecosystem is called the **water cycle**. In the water cycle, water evaporates from the Earth's surfaces, like the ocean, rivers, puddles, and even from plants; **evaporation** is the process of a liquid changing to a gas. Energy from the Sun drives the process of evaporation as solar radiation heats liquid water molecules and they change to the gas phase. Solar energy also evaporates water directly from the leaves of plants; it is called **transpiration**. (See the figure.)

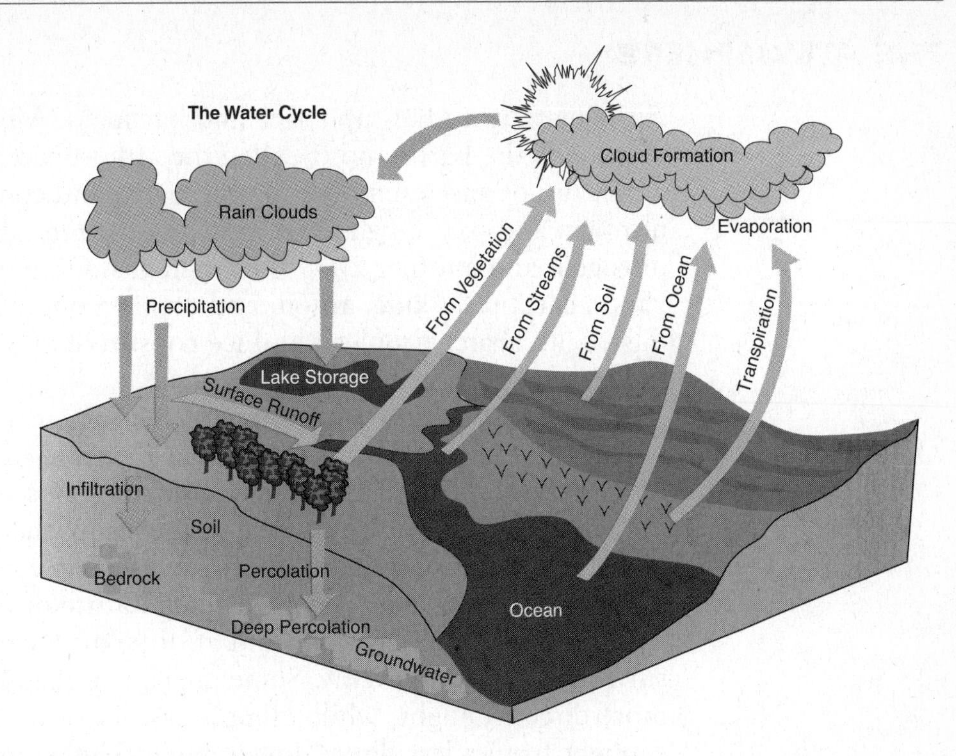

The gas form of water, or water vapor, rises and cools as it moves to higher elevations. As it cools, it condenses into liquid water particles in the air, forming clouds and fog. **Condensation** is the process in which a gas changes into a liquid. When the condensed water particles in clouds become too large to remain suspended, they fall to the surface as **precipitation.** Some clouds, though, do not produce precipitation. Precipitation can take the form of rain, snow, hail, or sleet depending on such factors as temperature and air pressure.

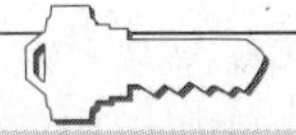

KEY CONCEPT

The processes of evaporation, condensation, precipitation, and runoff compose the water cycle.

The water then goes through the process of **runoff** whereby it moves across the land and collects in lakes, oceans, soil, and in rocks underground. It can then be evaporated, beginning the cycle again.

Example 1

What is the source of energy that drives the water cycle?

A. solar energy

B. global winds

C. Sun's gravity

D. Earth's gravity

See page 181 for the answer.

THE ATMOSPHERE

Air is matter; it takes up space and has mass. A blanket of air surrounds the Earth and is called the atmosphere. The **atmosphere** is a mixture of gases surrounding the Earth and consists mostly of nitrogen and oxygen gases. About 78% of the atmosphere is nitrogen, and another 20% is oxygen gas. Other gases such as water vapor, carbon dioxide, argon, and ozone along with particles like dust, salts, water droplets, and ice constitute only a tiny portion of the air around us.

The atmosphere extends about 100 kilometers above the surface of the Earth. But, the atmosphere has different properties at different elevations. The air is much denser closer to the surface of the Earth. As you move upward in the atmosphere, the air becomes less dense. This is due to the gravitational force of the Earth, which pulls the gases inward. Also, the temperature of the atmosphere varies at different elevations and at different locations. The Earth's surface is heated unevenly. Some areas, like the equator, receive more direct sunlight, while other areas, like the poles, receive sunlight from a less direct, lower angle. The temperature of the atmosphere is measured using a **thermometer**. A thermometer consists of a tube that contains mercury or alcohol. As the liquid is heated, it expands, rising up the tube; as it cools, it contracts. The two most commonly used scales for measuring temperature are Fahrenheit and Celsius. On the Fahrenheit scale, the freezing point of water is 32 degrees and the boiling point is 212 degrees. On the Celsius scale, the freezing point of water is 0 degrees and the boiling point is 100 degrees. (See the figure on page 143.)

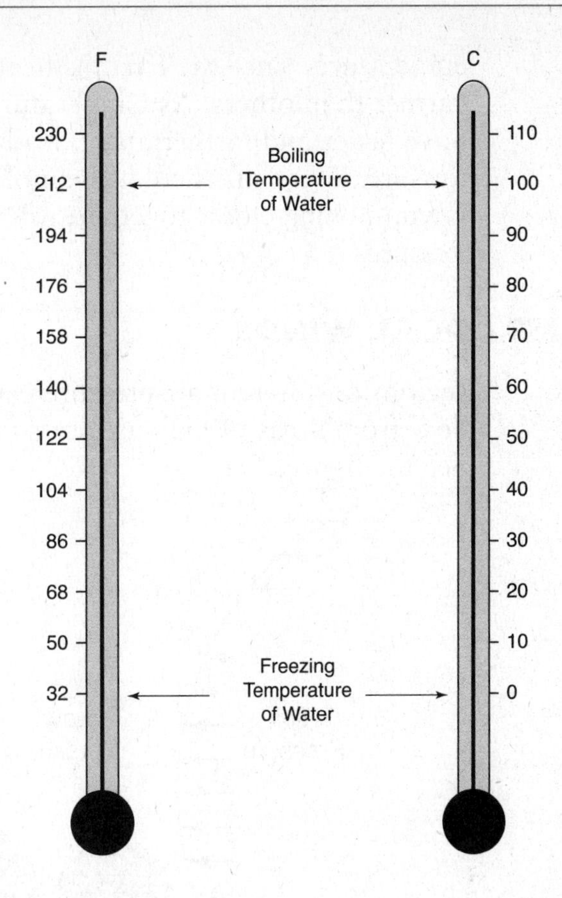

Example 2

The atmosphere is a mixture of gases. Which of the choices below places the gases in order of most abundant to least abundant in our atmosphere?

A. nitrogen gas, carbon dioxide gas, oxygen gas

B. oxygen gas, argon gas, nitrogen gas

C. nitrogen gas, oxygen gas, argon gas

D. carbon dioxide, nitrogen gas, oxygen gas

See page 181 for the answer.

The uneven heating affects not only the temperature of the atmosphere but also the air pressure. Gases in the air exert a pressure on all surfaces they contact, as the gas molecules collide with each other and the surface. On average, the atmosphere has an air pressure of about 14.7 pounds per square inch. Generally, when you travel higher in the atmosphere, the air pressure decreases. You may sense a change in air pressure when your ears pop in a plane or when driving up a mountain. The air pressure is also affected by air temperature. Since the Earth is heated unevenly, some areas are warmer than others. As temperature increases, the gas molecules move faster and farther apart, leading to a decreased, or lower, air pressure. When air is cold, the molecules release energy and slow down, moving closer together; this leads to increased, or higher, air pressure.

GLOBAL AND LOCAL WINDS

Regions of different air pressure cause the circulation of air. Winds blow from areas of high air pressure to regions of low air pressure. (See the figures.)

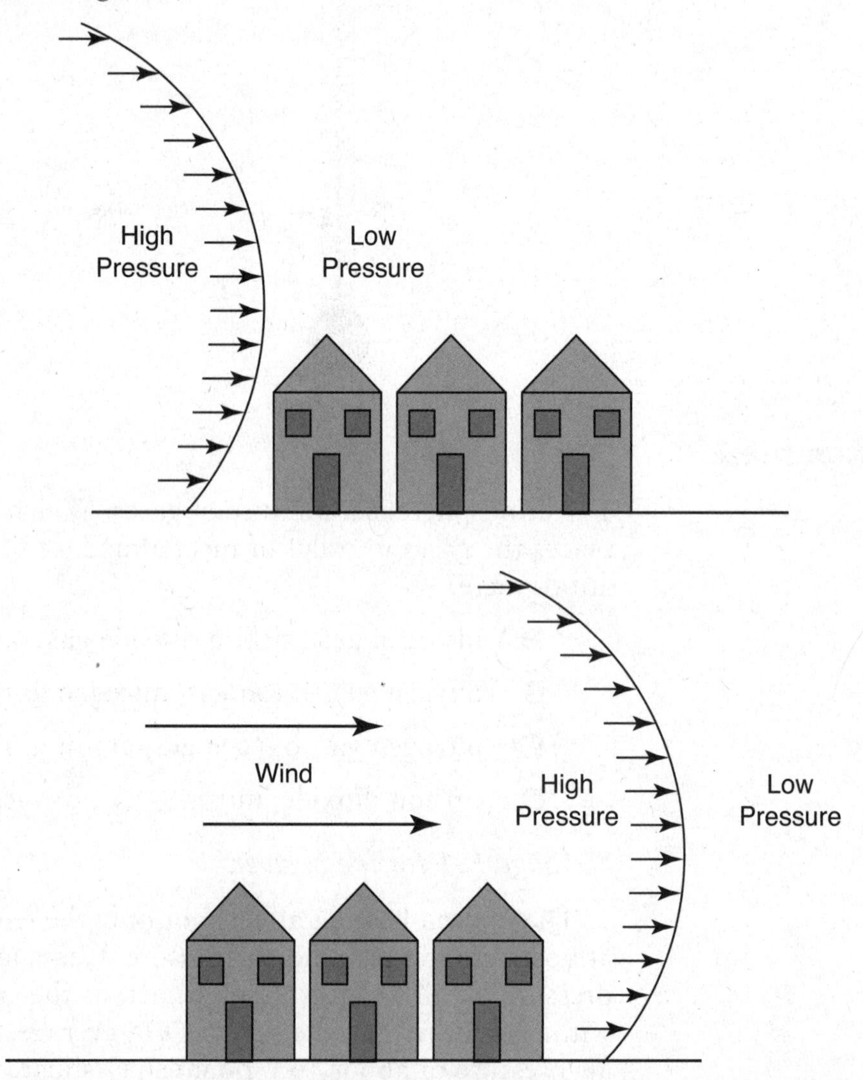

Some winds occur on a local scale. For example, land and sea breezes exist in regions where land and water meet. During the day, the land surface warms more easily than the water surface. As the air over the land warms and has a lower air pressure, it rises, and cooler, denser air (with a higher air pressure) from over the water's surface sweeps in. This is called a **sea breeze** because the air originated from over the sea. During the evening, the land cools faster than the seawater. The air over the sea is warmer and has a lower air pressure than the air over the land. The cooler air from the landside blows in the direction of the water. This is called a **land breeze** because the air originated from over the land. (See the figure.)

Land and Sea Breeze

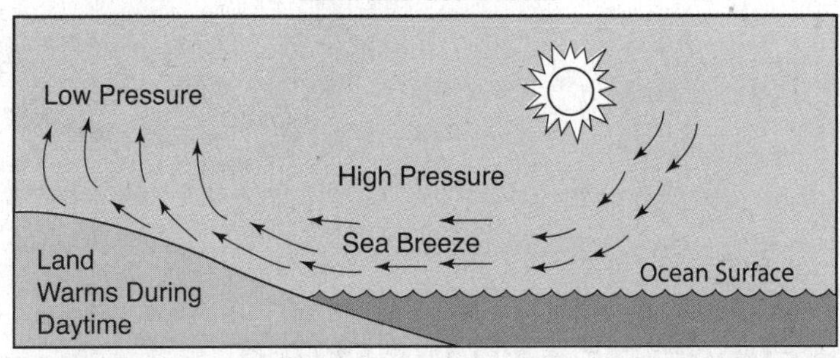

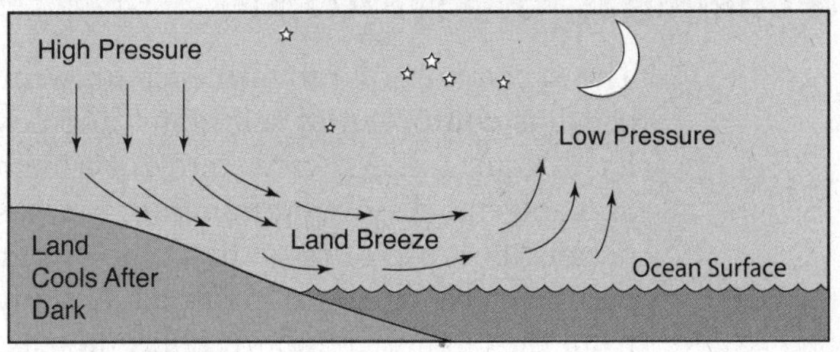

Differences in air pressure also cause winds on a global scale. Warm, less dense air over the equator is displaced by cooler air with a higher air pressure from adjacent regions. Air sinks at the poles, and the cold, denser air with high air pressure travels away from the poles in a southerly direction. These **global winds** circulate and mix the air. As you can observe in the figure, global winds are deflected, or curved, in a clockwise direction in the northern hemisphere and in a counterclockwise direction in the southern hemisphere. This curving of the path of global winds is called the **Coriolis effect** and is due to the Earth's rotation. These global patterns of air circulation even influence weather on a local scale. (See the figure on page 146.)

KEY CONCEPT

Winds blow from areas of high air pressure to regions of low air pressure.

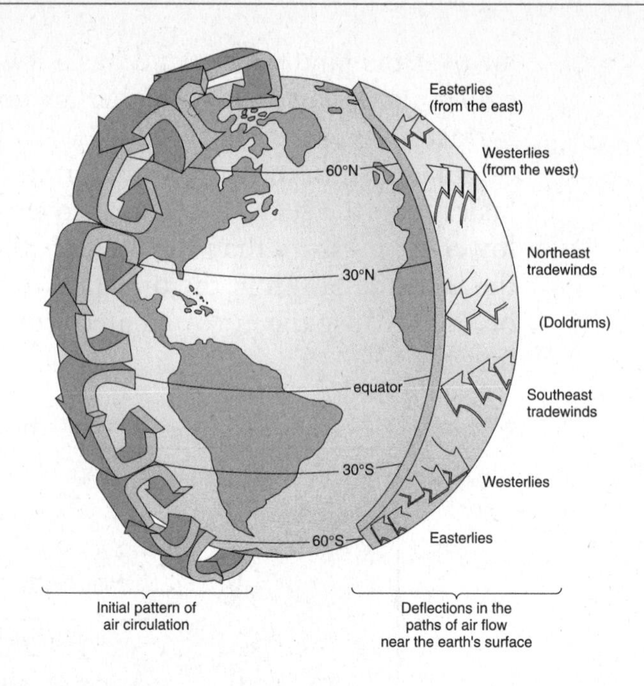

OCEAN CURRENTS AND WEATHER

The ocean has a large influence on weather patterns as well. The ocean is composed of salt water and covers roughly 70% of the Earth's surface. As water moves through the water cycle, water acts as a solvent, dissolving minerals and gases and carrying them to the oceans. Because of this, the ocean is salty. Just like the atmosphere, ocean waters circulate mainly due to uneven heating. Warm water from the equator region is continuously transported toward the poles. And, ocean water from the poles is transported toward the equator. These **ocean currents** are mostly driven by global winds. (See the figure.)

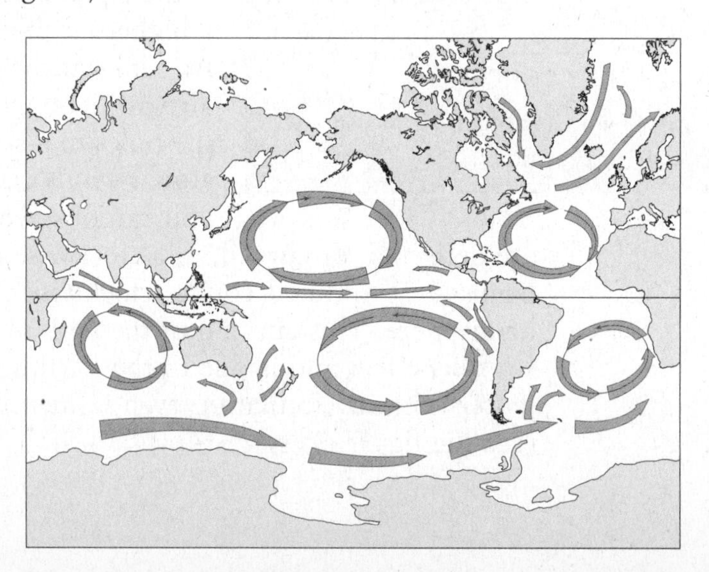

Example 3

If you were to sail using the currents off the coast of New Jersey, where would the currents most directly lead you?

A. Central America

B. Florida

C. Alaska

D. Europe

See page 181 for the answer.

The ocean affects the weather in many ways. Oceans hold a large amount of heat, and when air blows over the warm water, the temperature of the air tends to get warm as well. This has a major effect on climate in coastal areas. Also, when winds blow over the water, salt particles can get into the air. These particles are important in the water cycle because water condenses on these solids to form fog and clouds.

FORECASTING THE WEATHER

A variety of techniques and tools are used to forecast the weather. Thousands of scientists work around the globe to collect data used to make weather predictions. However, weather forecasting is not an exact science. Even with lots of data and the use of computer models, very often weather forecasts are incorrect.

Weather is the state of the atmosphere at a given place and time. Weather conditions are always changing and include temperature, moisture, winds, air pressure, and visibility. **Meteorologists**, or scientists who study weather, observe weather changes and record measurable quantities such as temperature, wind direction, wind speed, and air pressure. They look for patterns and make predictions based on previously observed weather conditions.

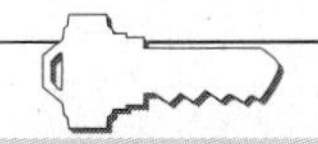

KEY CONCEPT

Weather is the state of the atmosphere at a given place and time and includes temperature, moisture, winds, air pressure, and visibility.

The measurement of air pressure is commonly used to forecast weather. A **barometer** is used to measure air pressure. If the barometric measurement indicates that air pressure is rising, then a simple forecast could be for pleasant weather. If the air pressure reading is lowering, then cloudy, rainy weather may be on its way. Clouds are also used to forecast weather. Various clouds are associated with specific weather changes and conditions. Cumulus and cirrus clouds are associated with fair weather, and cumulonimbus and nimbostratus clouds are associated with rain and snow. (See the figure.)

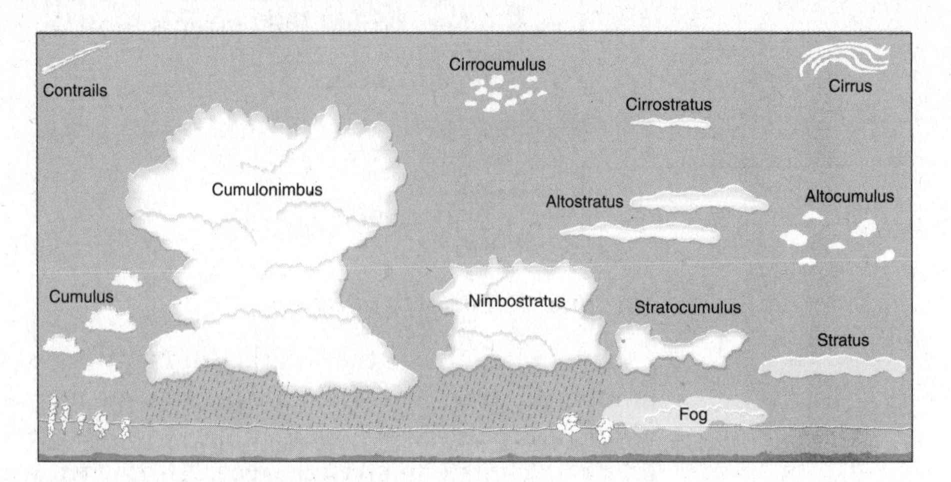

Forecasts are largely based on the movement of air masses. An **air mass** is a large region of air with uniform characteristics, such as temperature and pressure. The boundary between two air masses is called a **front**. A **cold front** forms as a cold air mass pushes against and under a warm air mass. Along the edge of the front, thunderstorms are common as the warm air is forced upward and moisture rapidly condenses, forming thick clouds. A **warm front** forms where a warm air mass pushes against and over a cold air mass. Since the warm air is less dense, it rides up over the cold air mass. The moisture in the warm air mass condenses and the air cools, causing thick, low clouds and precipitation. (See the figure.)

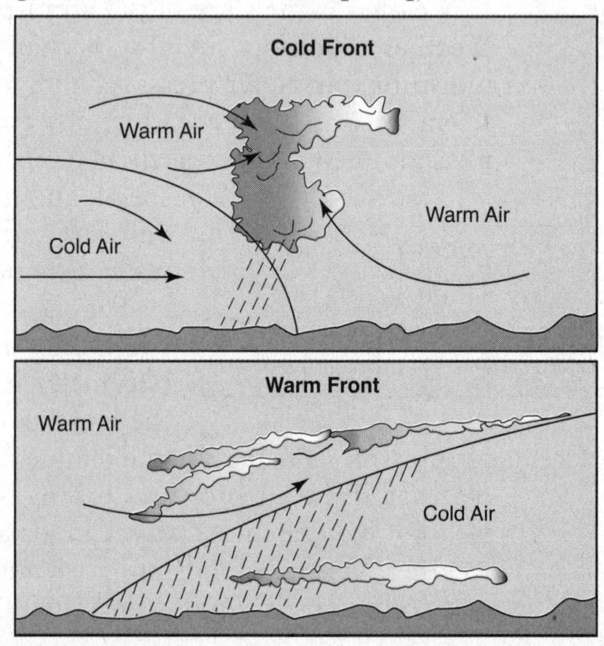

Meteorologists plot atmospheric data, like air pressure, on weather maps. A **weather map** can be used to represent weather conditions in a given area and predict the temperature and precipitation for several days. Various symbols are used on weather maps to represent fronts, areas of high and low pressure, and other weather conditions. (See the figures.)

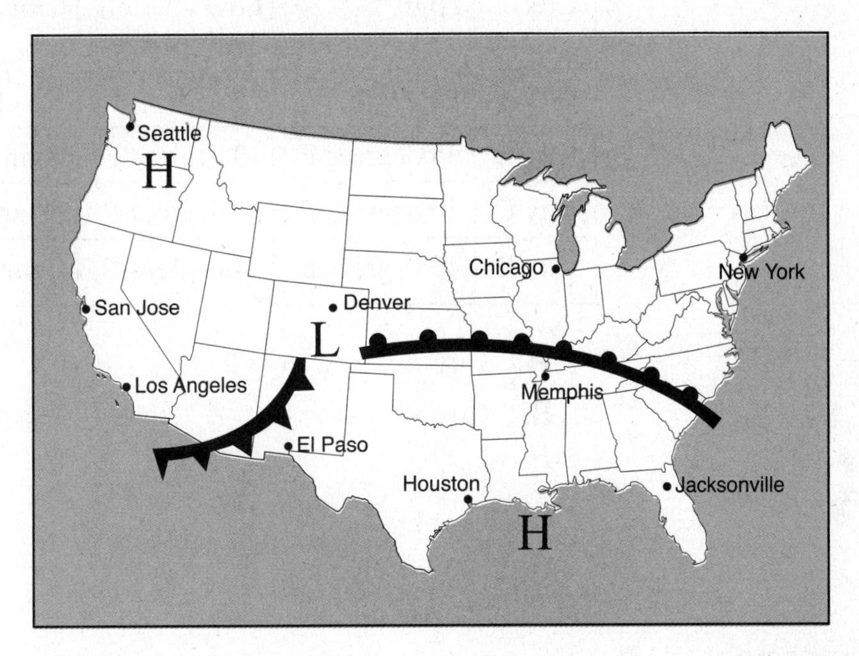

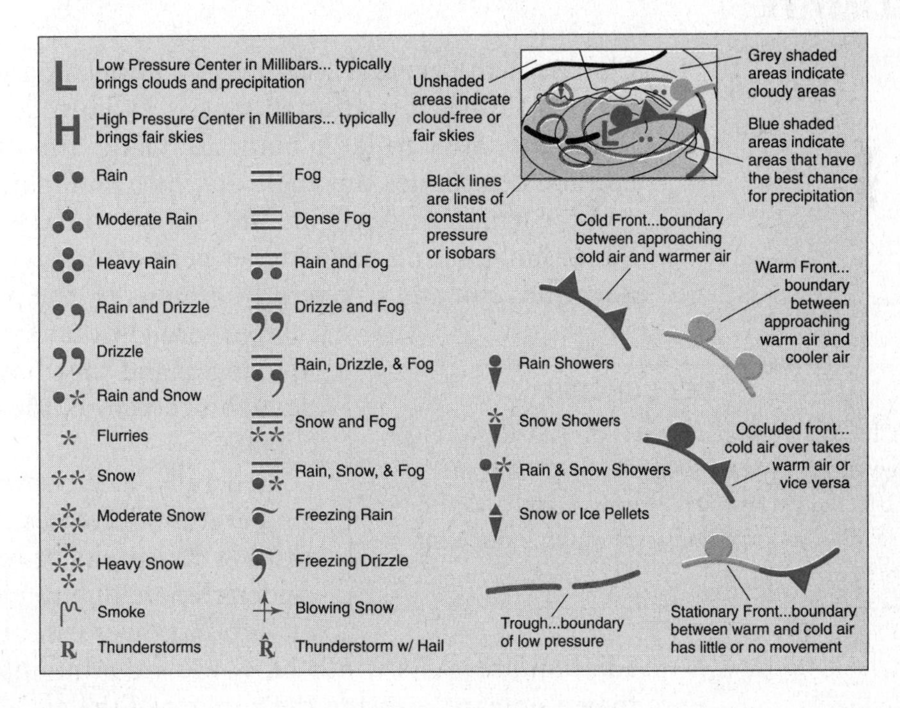

Example 4

The weather data in the chart was collected for four different cities on Tuesday. Which of these cities would be most likely to get snowfall on Wednesday?

City	High Temperature	Low Temperature	Cloud Type	Barometer Trend
City A	32 degrees F	25 degrees F	No clouds	Rising
City B	35 degrees F	10 degrees F	Nimbostratus	Falling
City C	50 degrees F	42 degrees F	Nimbostratus	Falling
City D	45 degrees F	34 degrees F	Cumulus	Rising

A. City A
B. City B
C. City C
D. City D

See page 181 for the answer.

CLIMATE

Climate is the general weather conditions in a given area over a long period of time. It is affected mostly by latitude, elevation, and location relative to large bodies of water and mountain ranges. The latitude determines the angle at which sunlight strikes the surface and the length of daylight. Areas close to the equator receive more direct sunlight and have longer periods of daylight; hence, they have a warmer climate. Conversely, regions by the poles of the Earth receive sunlight that strikes the Earth at a low angle, and they have shorter periods of daylight; therefore, they have a colder climate.

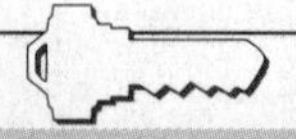

KEY CONCEPT

Climate is the general weather conditions in a given area over a long period of time and is affected by latitude, elevation, and location relative to large bodies of water and mountain ranges.

Generally, as elevation increases, the temperature decreases. Regions at a higher altitude and peaks of mountains will have a colder climate due to elevation. Also, mountain ranges can affect the climate of adjacent areas. As winds blow toward a mountain range, the air is forced upward, causing the cooling of the air and condensation of moisture. This side of the range will experience a cool, wet climate.

However, as the air moves over the range most of the moisture is lost. The air will warm as it descends in altitude. The other side of the range will experience a warm, dry climate.

The climate of a region is also affected by its location relative to water. Regions that are inland and far from large bodies of water usually have a dry climate. Areas that are near a large body of water are usually cold in the winter and warm in the summer because the body of water helps to moderate the temperature of the land.

Regional climates are classified by the levels of moisture, overall temperature, and type of vegetation. Refer to the figure for a summary of major climates zones of the Earth.

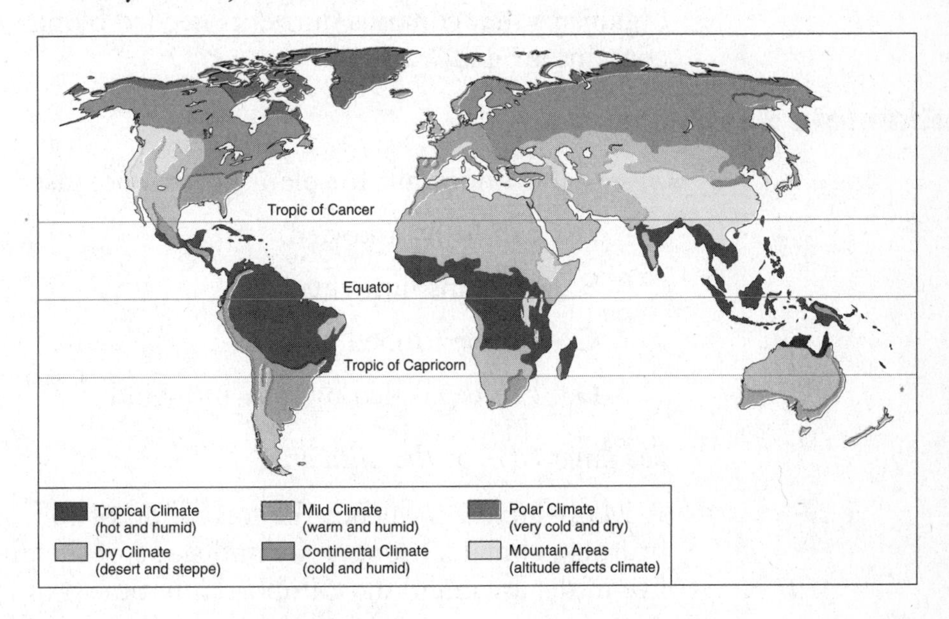

SOIL, MINERALS, AND ROCKS

After reading this section you should be able to:

- Recognize that both soil and rocks are made of several substances or minerals.
- Describe the characteristics of the three main classifications of rocks.
- Summarize the processes involved in the rock cycle.

SOIL

In the previous section, we reviewed the interactions between the atmosphere and the hydrosphere, or water systems on Earth. Now

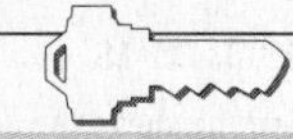

KEY CONCEPT

Soil consists of weathered rock, gravel, sand, silt, clay, and humus.

we will explore the geosphere, or the solid part of the Earth, which mostly consists of rock and soil. The outermost layer of the Earth's surface consists of soil. Relative to the size of the Earth, the layer of soil is extremely thin. Yet, we are incredibly dependent on it. Soil consists of weathered rocks and decomposed organic material from dead plants, animals, and bacteria. It can take up to 1,000 years for 1 centimeter of soil to form. The main components of soil are weathered rock, gravel, sand, silt, clay, and humus. Humus is dark, moist organic material composed of decomposed plants and organisms that contains nutrients needed by plants. Soil also contains air and water.

Example 5

Soil is important for plant growth because it

A. can be transported.

B. contains nutrients.

C. can be formed quickly.

D. can be eroded by rain and wind.

See page 181 for the answer.

Many factors influence the formation of soil, such as the bedrock, climate, topography and slope of the area, organisms that live in the area, and most importantly time. For example, some rock types break down easier than others. Quartzite, for instance, is more resistant than shale. Given similar climate conditions, soil will form a thicker layer over the shale rather than the quartzite. (See the figure.)

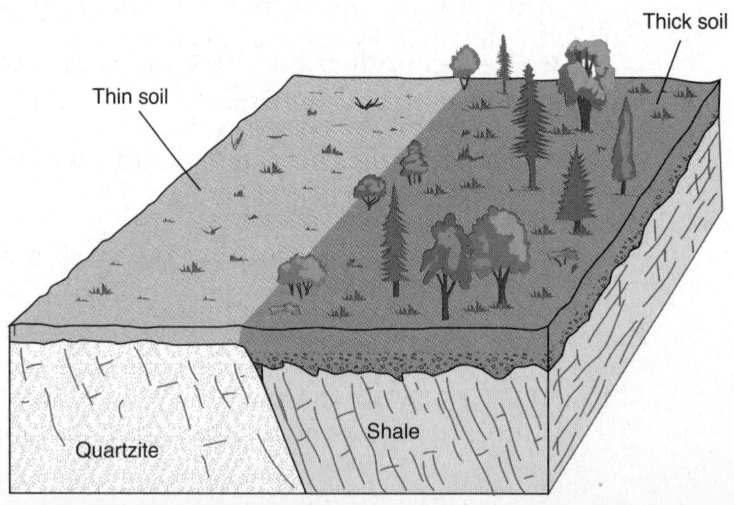

For soil to form, rock must weather or break down into smaller pieces. Shale weathers more easily than quartzite. The rock continues to break down into smaller and smaller pieces through physical weathering and chemical reactions. Plants begin to grow, and organic material accumulates in the soil. Organisms, like earthworms and insects, live in the soil and deposit more organic material. As time passes, layers form, each having a different chemical composition and texture. The layers are also called **horizons** and are named Horizon O, A, B, C, and R. (See the figure.)

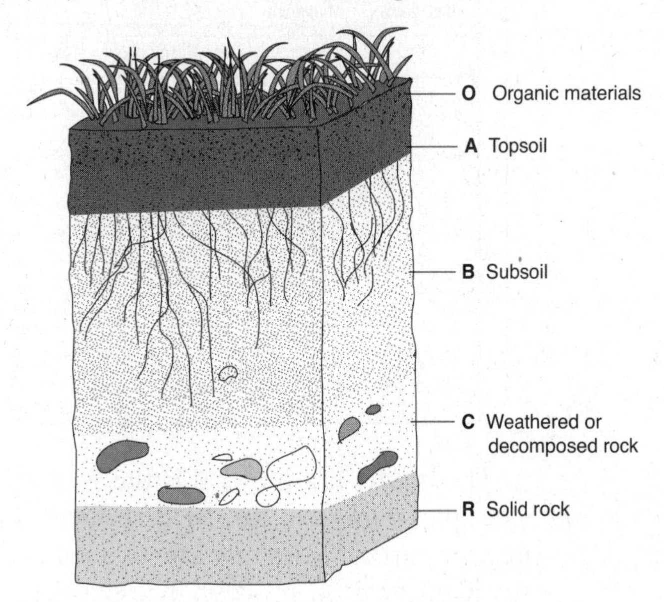

ROCKS AND MINERALS

Soils form from weathered rock and organic materials. But, what are rocks made of then? Rocks and soils are made of several substances or minerals. A **rock** is made of one or more minerals. **Minerals** are naturally occurring, inorganic substances that have a definite chemical composition. They are inorganic in that they do not come from living things. And, they have a definite chemical composition; minerals are either elements or compounds, which means that a chemical symbol or formula can be used to represent them. Some common minerals that compose many rocks are feldspar, quartz, and mica.

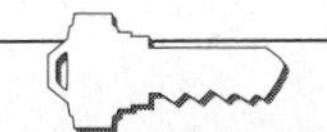

KEY CONCEPT

A rock is made of one or more minerals. Minerals are naturally occurring, inorganic substances that have a definite chemical composition.

The physical properties such as color, hardness, luster, and streak can be used to help identify minerals. The color of a mineral is usually the first observation made; however, using this property alone is an unreliable method of identification. Some minerals, like quartz, can be found in a range of colors from yellow to purple. The **hardness** of a mineral is quite useful for identification. A scale called the **Moh's hardness scale** can be used to determine the mineral's resistance to being scratched. Talc, for example, is very soft and can be scratched with a fingernail. Diamond, on the other hand, is very hard and can only be scratched by another diamond.

Moh's Hardness Scale

Hardness	Mineral	Description
10	Diamond	Scratches all minerals. Can only be scratched by another diamond.
9	Corundum	Scratches all minerals but diamond.
8	Topaz	
7	Quartz	Scratches glass.
6	Orthoclase	Can be scratched with a steel file.
5	Apatite	
4	Fluorite	
3	Calcite	Can be scratched with a penny.
2	Gypsum	
1	Talc	Can be scratched with a fingernail.

The **luster** of a mineral is the way that light interacts with it. For example, a mineral can be described as metallic, glassy, pearly, waxy, oily, or earthy. Minerals are also rubbed on a hard surface as part of the identification process. The **streak**, or color of the powder left behind after being rubbed across a hard surface, can be used to identify the mineral. Some minerals, such as calcite, react with acid. When hydrochloric acid is dropped on calcite, the mineral bubbles vigorously.

Minerals are the building blocks of rocks. Every rock contains one or more minerals. Rocks are classified into three basic groups, not by the minerals they contain, but by how the rock formed. Rocks can be classified as igneous, sedimentary, or metamorphic.

IGNEOUS ROCKS

Hot, molten rock material called **magma** is found deep below the Earth's surface. **Igneous rocks** form when this liquid magma cools and becomes a solid. Sometimes igneous rocks form on the surface of Earth, and other times they form within the Earth. When magma reaches the surface it is called **lava**. Lava can erupt at volcanoes or beneath the ocean at midocean ridges where new ocean floor is created. **Basalt** is a type of igneous rock that usually forms at midocean ridges and composes the ocean floor. When magma

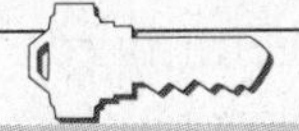

KEY CONCEPT

Igneous rocks form when magma cools and solidifies.

reaches the surface of the Earth, such as at volcanoes or at midocean ridges, the lava cools and solidifies very quickly. However, when magma cools below the surface of the Earth, it solidifies very slowly. When it solidifies slowly, large mineral crystals can form giving the rock a very coarse-grained appearance. **Granite** is a type of igneous rock that forms slowly below the Earth's surface; it can contain crystals of the minerals quartz, feldspar, and plagioclase.

SEDIMENTARY ROCKS

The weathering of rocks forms **sediments**, like gravel, sand, silt, and clay. These sediments are transported by water, wind, or ice and accumulate in layers. Over time, the layers build up; air and water are squeezed out due to the weight above. The sediments undergo **lithification** and are converted into **sedimentary rocks**. Lithification is the process in which the grains of sediment are compacted and cemented together, forming sedimentary rocks. Sandstone and shale are examples of common sedimentary rocks.

KEY CONCEPT

Sedimentary rocks are formed from the lithification of sediments, like sand, silt, and clay, and they often contain fossilized remains of organisms.

Sedimentary rocks are important because they can contain the fossilized remains of living organisms. Some sedimentary rocks, such as limestone, chalk, and coal, are even composed entirely of the remains of living organisms.

Example 6

Which part of a dinosaur is most likely to be preserved as a fossil?

A. stomach

B. brain

C. muscle

D. bone

See page 181 for the answer.

METAMORPHIC ROCKS

Metamorphic rocks form deep below the surface of the Earth where heat and pressure change the preexisting solid rocks. The preexisting rocks can be igneous, sedimentary, or metamorphic. The intense heat and pressure cause **metamorphism**, physical and chemical

KEY CONCEPT

Metamorphic rocks form deep below the surface of the Earth whereby heat and pressure change the preexisting solid rocks.

changes in the rock, which can even make them appear folded. Metamorphism can turn limestone to marble, shale to slate, and slate to schist. Common metamorphic rocks are marble, gneiss, and quartzite.

THE ROCK CYCLE

The Earth is 4.6 billion years old! During this immense time period, the solid rock that composes the outermost layer of our planet has undergone tremendous change. The rock cycle is a fundamental concept in geology, the study of the Earth. Because of weathering, erosion, and the movement of the Earth's crust, rocks are constantly undergoing change. Rocks are formed, changed, melted, and reformed through the rock cycle. Rocks at the surface are being weathered and eroded. Sediments are being layered and lithified, and then perhaps compacted further through metamorphism; they may even be melted and solidified into igneous rock again; this is the rock cycle. (See the figure.)

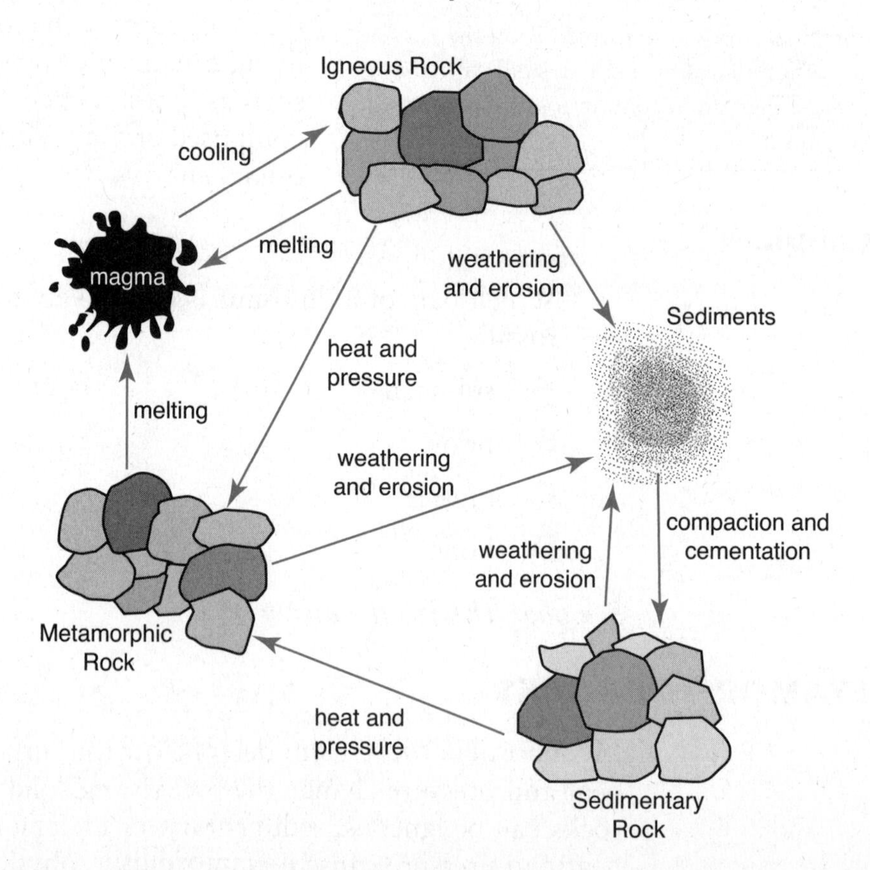

Igneous rocks can be melted and solidified again; they can be weathered and cemented into sedimentary rocks, or they can be changed through heat and temperature into metamorphic rocks. Sedimentary rocks could be weathered into sediments again, melted and solidified, or changed through heat and temperature. And, metamorphic rocks can be weathered and eroded, and then compacted into sedimentary rocks, or they can be melted and cooled, forming igneous rocks. The rock cycle is the recycling of Earth materials and is driven by the water cycle and plate tectonics.

Example 7

Which of the following changes is NOT part of the rock cycle?

A. Weathering and erosion convert metamorphic rocks into igneous rocks.

B. Heat and pressure convert igneous rocks into metamorphic rocks.

C. Weathering and erosion convert igneous rocks into sediments.

D. Heat and pressure convert sedimentary rock into metamorphic rocks.

See page 181 for the answer.

THE DYNAMIC EARTH

After reading this section you should be able to:

- Explain how Earth's landforms are created through constructive and destructive processes.
- Recognize that moving water, wind, and ice continually shape the Earth's surface.
- Understand that fossils provide evidence about the plants and animals that lived long ago and the nature of the environment at that time.

THE EARTH'S INTERIOR

In the last section, we learned about the materials that make up the outermost layer of the Earth. The soil and solid rock make up the relatively thin crust of the Earth. Beneath the crust are layers made of different materials with varying properties. The lithosphere is composed of the crust and the uppermost solid portion of the next layer, the mantle. The **mantle** is the thickest layer of the Earth; it is hot, and although it is solid, it slowly flows and circulates by convection currents. Within the mantle, but below the lithosphere, is a layer of the mantle called the **asthenosphere**. The lithosphere floats on the asthenosphere, which is made of partially melted, deformable rock. Below the mantle is the core. The **outer core** is believed to be dense, liquid iron and nickel. The circulation of this liquid metal generates the Earth's magnetic field. The **inner core** is believed to be extremely hot, dense, solid iron. (See the figure.)

Earth's Interior

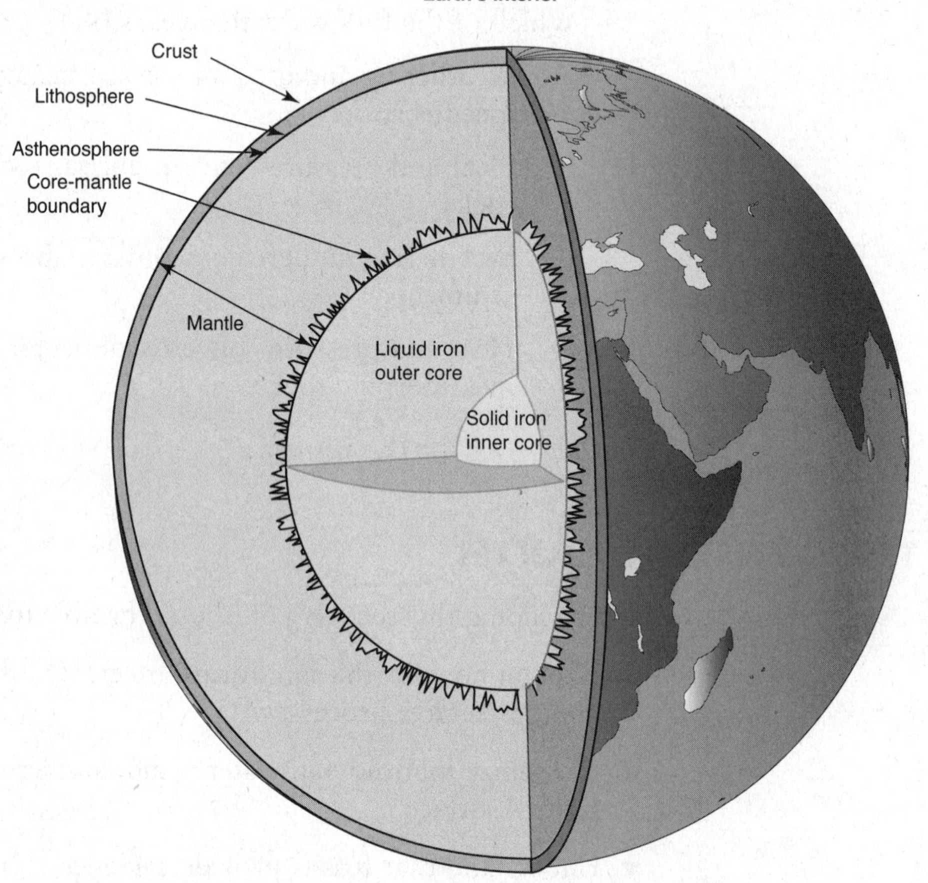

Scientists infer the composition and behavior of the materials in the Earth's interior by observing the seismic waves released by earthquakes, by the effect of the Earth's gravitational force on other objects, and by attempting to measure the density of the planet by determining the Earth's mass and volume. The Earth's radius is around 4,000 miles, yet we have only drilled down

> **KEY CONCEPT**
> Earth's interior is layered with a lithosphere, hot, convecting mantle, and dense, metallic core.

roughly seven miles. Despite much effort, scientists have yet to reach the mantle. Therefore, our understanding of the Earth's interior is limited to inferences based on observations.

Example 8

What is the only layer in the interior of the Earth that is composed of liquid?

A. crust

B. mantle

C. outer core

D. inner core

See page 181 for the answer.

PLATE TECTONICS

Plate tectonics is a theory stating that the lithosphere is broken up into plates that float independently over the asthenosphere at a rate of centimeters per year. There are about 12 large plates and several smaller ones; the plates contain both continental and oceanic crust. Considering the age of the Earth, 4.6 billion years, the plates have traveled tremendous distances. (See the figure.)

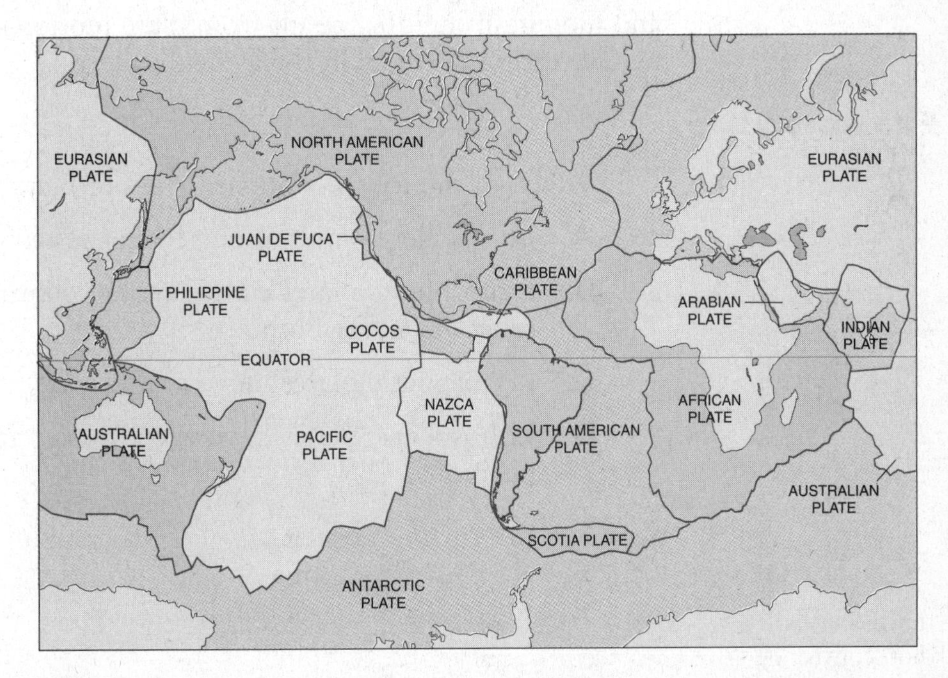

Plate tectonics is a relatively new geologic theory. Previously, scientists believed in the concept of **continental drift**, whereby continents were once one large supercontinent, Gondwana, and have since drifted through the oceans to their current locations. Plate tectonics transformed this theory by suggesting that the continents are parts of larger plates containing *both* continental crust and oceanic crust. This newer theory also incorporated the theory of **sea floor spreading**. The theory of sea floor spreading suggests that oceanic crust is created through volcanic activity along ocean ridge systems. (See the figure.)

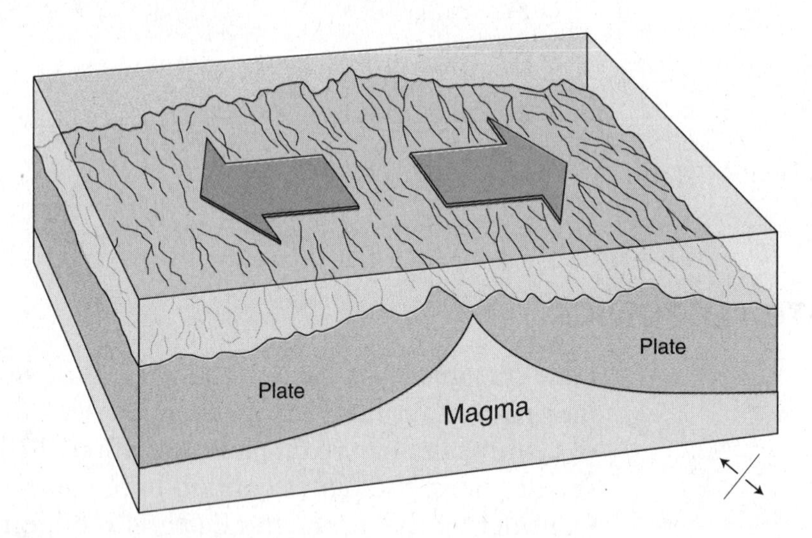

Major geological events, such as earthquakes, volcanic eruptions, and mountain building result from plate motions, which are driven by convection currents in the asthenosphere.

Example 9

Which of the follow statements about plate tectonics is true?

A. Lithosphere plates are no longer moving.

B. Lithospheric plates contain either continental or oceanic crust but never both.

C. Lithospheric plates move at a rate of centimeters per year.

D. Lithospheric plates have never moved from their current positions.

See page 181 for the answer.

EARTHQUAKES

Earthquakes often occur at the boundaries between lithospheric plates. Actually, scientists have used the locations of both earthquakes and volcanoes to help determine the locations of plate edges. The map below indicates the locations of earthquakes that occurred in the last 30 days (since accessing the U.S. Geological Survey web site). The earthquakes are primarily located by plate edges.

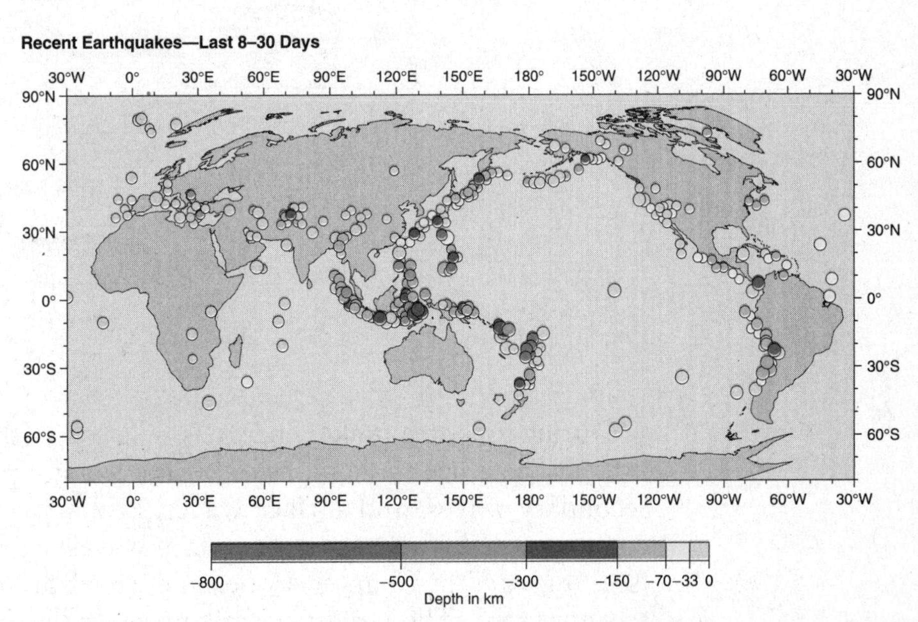

At some plate boundaries, rocks collide or slide past each other. If the walls of rock move relative to one another at a fracture, then it is known as a **fault**. At faults, the pressure accumulates as the rocks try to move. An earthquake occurs when there is movement along a fault and the accumulated pressure is released as energy. Earthquakes occur at all three basic types of faults: normal, reverse (or thrust), and strike-slip. Refer to the figure on page 161, which illustrates these faults.

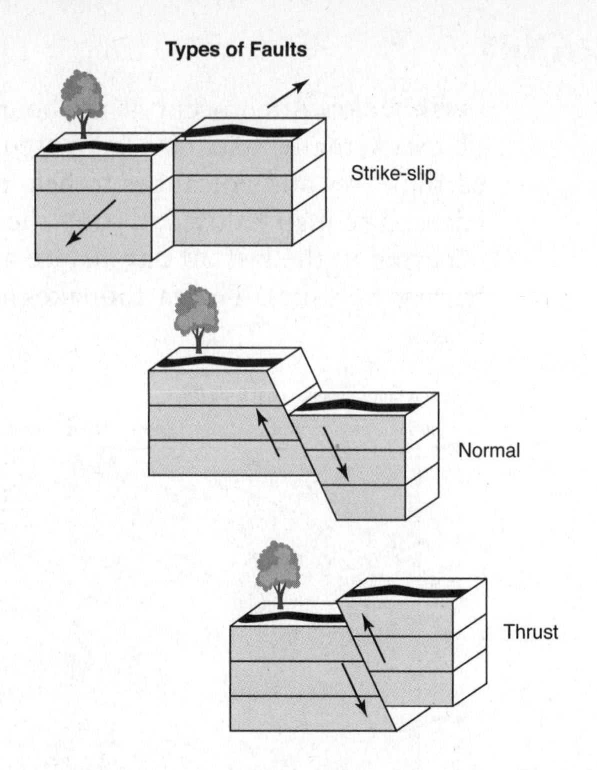

During an earthquake, energy is released in the form of seismic waves. There are three types of seismic waves: primary waves, secondary waves, and surface waves. Primary waves are compressional waves, and secondary waves are longitudinal waves (see "Energy" in Chapter 4). Seismic waves are recorded using seismographs. The largest seismic waves collected using the seismograph for a particular earthquake are then used to determine the magnitude of that earthquake on the Richter scale. (See the figure and table.)

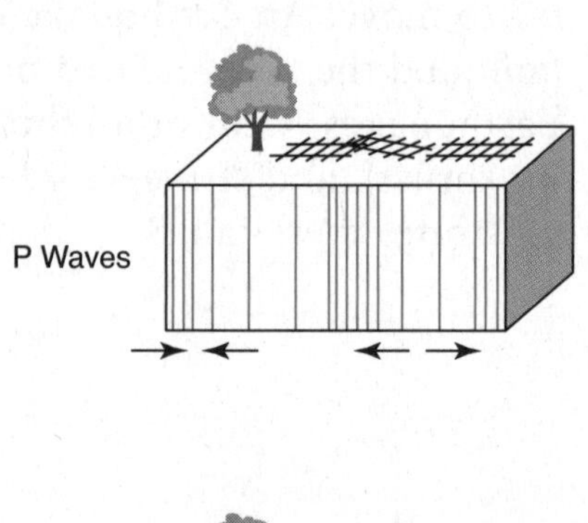

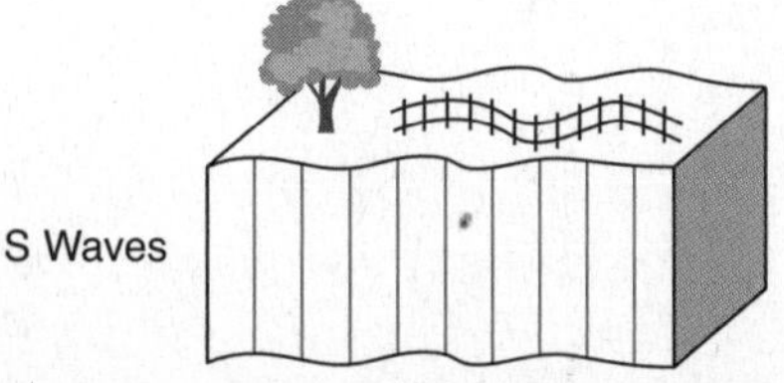

Richter Magnitudes	Description	Earthquake Effects	Frequency of Occurrence
Less than 2.0	Micro	Microearthquakes, not felt.	About 8,000 per day
2.0–2.9	Minor	Generally not felt, but recorded.	About 1,000 per day
3.0–3.9	Minor	Often felt, but rarely causes damage.	49,000 per year (est.)
4.0–4.9	Light	Noticeable shaking of indoor items, rattling noises. Significant damage unlikely.	6,200 per year (est.)
5.0–5.9	Moderate	Can cause major damage to poorly constructed buildings over small regions. At most slight damage to well-designed buildings.	800 per year
6.0–6.9	Strong	Can be destructive in areas up to about 100 miles across in populated areas.	120 per year
7.0–7.9	Major	Can cause serious damage over larger areas.	18 per year
8.0–8.9	Great	Can cause serious damage in areas several hundred miles across.	1 per year
9.0–9.9	Great	Devastating in areas thousands of miles across.	1 per 20 years
10.0+	Great	Never recorded.	Extremely rare (unknown)

PLATE BOUNDARIES

New features on Earth's surface are formed as a result of dynamic forces. Some of the forces are constructive and build up the crust through volcanoes, sediment deposition, and convergent plate boundaries. Other forces, like weather and erosion, are destructive, breaking down the surface materials.

Plate tectonics is one of the forces that transforms the surface of the Earth. At plate boundaries, events such as earthquakes and

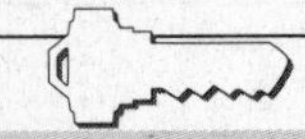

KEY CONCEPT

Earthquakes usually occur at plate boundaries, where there is movement along a fault. The accumulated energy is released as seismic waves.

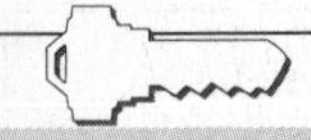

KEY CONCEPT

Plate tectonics is a constructive force, building new landforms along plate boundaries.

volcanoes occur, and new landforms are created. Earthquakes and volcanic eruptions can abruptly change the surface of the Earth. Volcanic eruptions can uplift the Earth's surface, forming mountains that spew lava.

Lithosphere plates converge, diverge, and slide past each other. At **convergent plate boundaries**, plates collide; at **divergent plate boundaries**, plates move apart; and, at **transform plate boundaries**, plates slide past each other. The interactions between plates at these boundaries in affected by the density of the rocks in the plates. Oceanic crust is usually made of the rock basalt, which is denser than continental crust, which is usually made of granite. (See the figure below.)

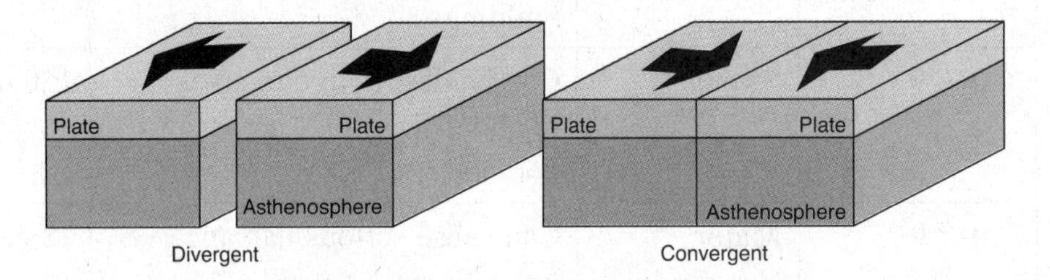

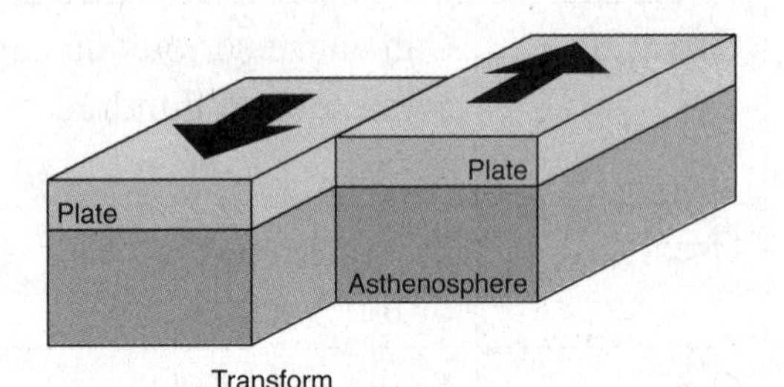

Various landforms are created at plate boundaries. (See the figure on page 165.)

In the top picture, two oceanic plates are diverging. At this boundary a midocean ridge and rift valley form; new oceanic crust is created as the plates move apart. The second picture down illustrates a convergent plate boundary. The oceanic crust is denser, so it **subducts**, or sinks beneath the continental crust. As the oceanic crust subducts it pushes up the continental crust forming a mountain range. The third picture down also illustrates a convergent boundary that is between two continental plates. Since they have similar densities, tall mountain ranges form at these types of boundaries. In the bottom picture, at a transform plate boundary, the rocks will grind past each other causing a breakdown in the rock. Eventually a small, shallow trench will form at the boundary and the bedrock will be weakened.

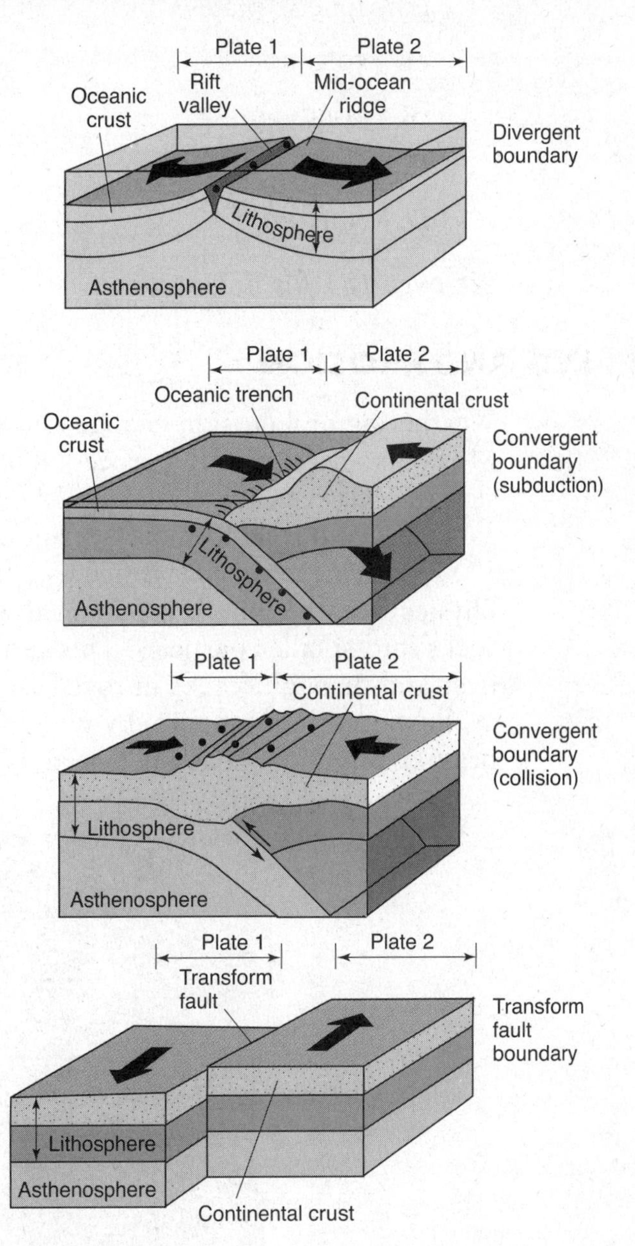

Example 10

Which landform is most likely created by the movement of lithospheric plates?

A. volcano

B. pond

C. river

D. canyon

See page 181 for the answer.

WEATHERING AND EROSION

Weathering and erosion change and wear down the Earth's surface. Most changes due to the processes of weathering and erosion occur slowly. However, some, like landslides, occur rapidly. Features such as valleys and soils form as a result of these destructive processes.

Weathering can be classified as either physical or chemical. **Physical weathering**, or **mechanical weathering**, is disintegration of rocks into smaller particles. This can be caused by water rushing over and through cracks in rocks gradually breaking off pieces and, as illustrated in the figure, by water freezing in the cracks and heaving pieces apart as the frozen water expands.

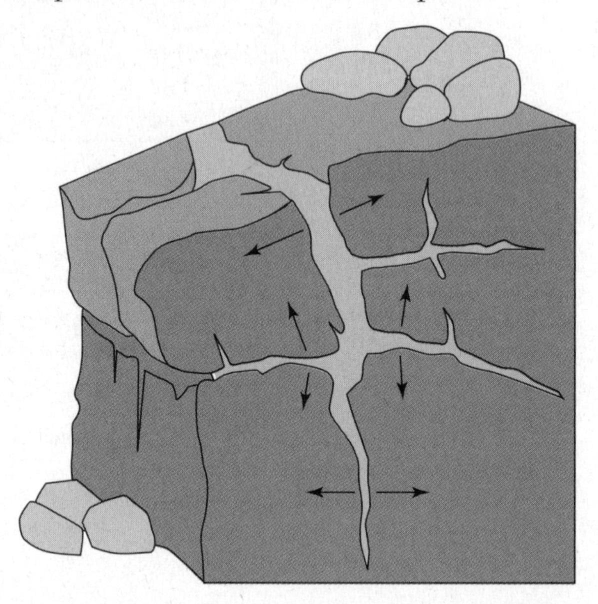

Chemical weathering usually involves a breakdown, or change, in the chemical composition of rocks. Chemicals, dissolved in rainwater, react with some rocks, causing chemical changes and producing new substances, like clay. Some organisms, such as mosses and lichens, may live on rocks and release substances that react with the rocks.

After rocks are weathered, erosion may occur, moving the soil and rocks to new locations. Wind, water, ice, and glaciers can erode, or move sediments to other locations. Sometimes erosion happens quickly as in areas prone to landslides. In other areas, streams and rivers slowly erode the materials on their banks and deposit the sediments downstream. Water has the capacity to lift heavy and large sediments. In a river, the faster the water travels, the larger the sediments it can erode. Glaciers gouge rocks and carry sediment suspended in the ice. When the glacier melts, all of the rock and sediment wash out with the melt water. Winds will erode soil and sediments in dry areas with little vegetation.

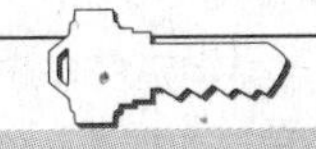

KEY CONCEPT

Weathering and erosion are destructive forces wearing down the Earth's surface.

USING FOSSILS TO DETERMINE EARTH'S HISTORY

Scientists gather evidence to determine the 4.6 billion year history of the Earth. The Earth processes we see today, including weathering, erosion, and movement of lithospheric plates, are similar to those that occurred in the past. During Earth's history, occasional catastrophes, such as the impact of an asteroid or comet, occurred. These catastrophic events influenced our planet, too.

The main source of evidence that provides clues to Earth's history are layers of sedimentary rocks and the fossils contained within them. Fossils provide important evidence of how life and environmental conditions have changed throughout Earth's history. Fossils provide evidence about the plants and animals that lived long ago and the nature of the environment at that time. They can be used to confirm the age, history, changing life forms, and geology of Earth.

Fossils usually consist of mineralized hard parts of organisms. Sometimes imprints of leaves or shells are found, too. Rarely are the soft parts of organism preserved. After the organism is covered in sediments, the soft parts decompose and are usually not preserved, but the bones and other hard parts are hardened and preserved.

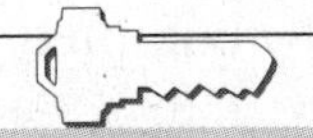

KEY CONCEPT

Sedimentary rocks and the fossils within them provide important evidence of how life and environmental conditions have changed throughout Earth's history.

When analyzing layers of sedimentary rocks and the fossils within them, scientists use the law of superposition to relatively date the rocks and fossils. The law of superposition states that sedimentary layers are deposited in a time sequence. Older layers are on the bottom and younger layers are on the top. Scientists also compare rocks and fossils found in various locations within similar regions. The law of lateral continuity states that layers of sedimentary rock extend laterally in all directions. In other words, rocks that are now separated by a building, roadway, or valley may actually be part of the same layer of sediment.

EARTH IN SPACE

After reading this section you should be able to:

- Interpret a diagram that shows how the positions of the Earth, Sun, and Moon affect the tides, phases of the moon, and/or eclipses.
- Explain how the motions of the Earth, Sun, and Moon define units of time including days, months, and years.
- Describe the physical characteristics of the planets and other objects within the solar system and compare Earth to the rest of the planets.

EARTH-MOON-SUN SYSTEM

The Earth is the third planet from the Sun in our solar system. Orbiting the Earth is our Moon. Even though the Sun is about 93 million miles away from the Earth, the Sun affects Earth's processes. The Earth, Moon, and Sun compose a system that affects tides, phases of the moon, and eclipses.

Time is related to motion within the Earth-Moon-Sun system. One day on our planet generally consists of the length of time for the Earth to complete one rotation; the Earth rotates, or spins on its imaginary axis, once every 24 hours. One month is roughly equivalent to 29 days, or the length of time for the Moon to orbit around the Earth. One year is 365 days, or the length of time for the Earth to complete one revolution around the Sun.

KEY CONCEPT

Motions within the Earth-Moon-Sun system that affects seasons, tides, phases of the moon, and eclipses.

Example 11

How long does it take the Earth to complete one revolution around the Sun?

A. one year

B. one season

C. one month

D. one day

See page 182 for the answer.

The Earth is tilted at an angle of 23 degrees. As the Earth revolves around the Sun during the course of the year, changes occur; the amount of daylight changes, and other seasonal weather patterns evolve. These changes are due to the position of the Earth relative to the Sun, the latitude, the tilt of the Earth, and the angle of the sunlight as it strikes regions of our planet. In the summer, the northern hemisphere is tilted toward the Sun and receives direct sunlight for a longer amount of time per day. In the winter, the northern hemisphere is tilted away from the Sun and receives sunlight from a lower angle for a shorter amount of time per day. (See the figure.)

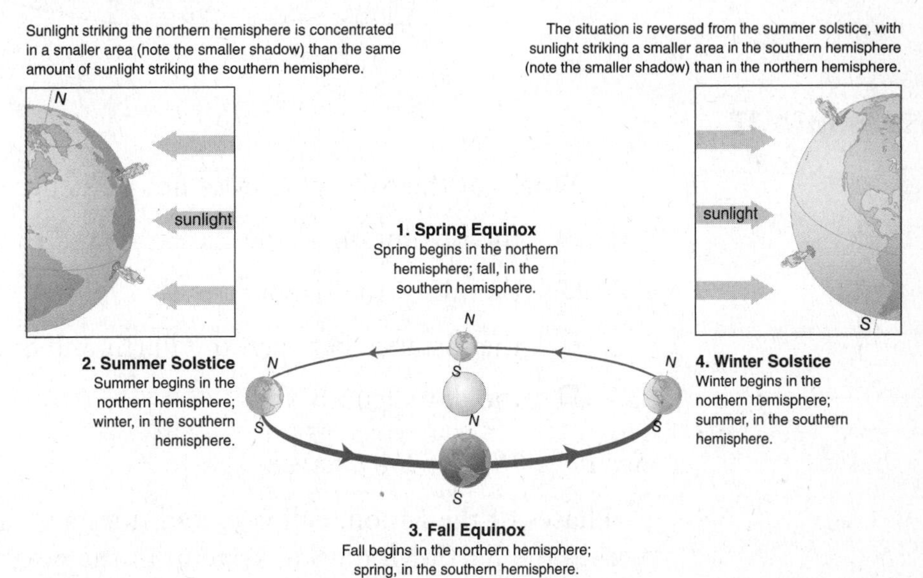

The figure on the next page shows the path of the Sun relative to a house in New Jersey on both the first day of summer and the first day of winter. During the summer, the path of the Sun travels directly overhead, efficiently heating the surface. However, during the winter, the Sun's path is lower in the sky. The Earth's surface is

KEY CONCEPT

The position of the Earth as it revolves around the Sun and, consequently, the amount of direct sunlight an area receives affects seasonal weather patterns.

not heated as much when the Sun's rays strike the surface from a lower angle. Regions close to the equator do not experience seasonal changes as dramatic as those regions in latitudes north and south of the equator. The seasonal changes in these areas are again due to variations in the amount of the Sun's energy hitting the surface, which is due to the tilt of the Earth and the length of the day.

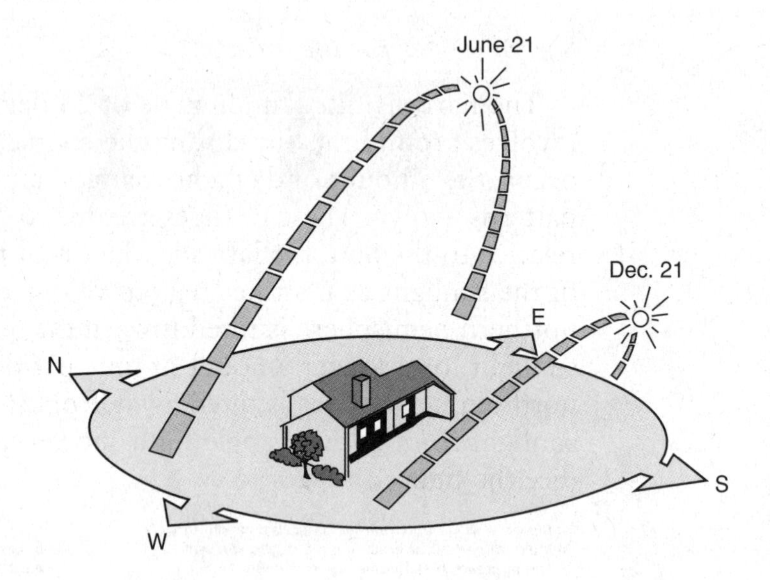

Example 12

Which of the following does not influence seasonal changes?

A. the revolution of the Earth around the Sun

B. the tilt of the Earth

C. the distance between the Earth and Sun

D. the day's length

See page 182 for the answer.

Phases of the Moon, eclipses, and tides are caused by motion within the Earth-Moon-Sun system. In the next few pages we will review the positions of Earth, Moon, and Sun during these phenomena.

PHASES OF THE MOON

The Moon revolves around the Earth. As the Moon revolves around the Earth, it is simultaneously rotating at the same speed that it revolves. This causes an interesting phenomenon; we can only observe one side of the Moon from Earth. Even though we can only see one side of the Moon, the moon appears to change. At various times of the month, different portions of the Moon appear to be illuminated as displayed in the figure; this is called the phases of the Moon. (See the figure.)

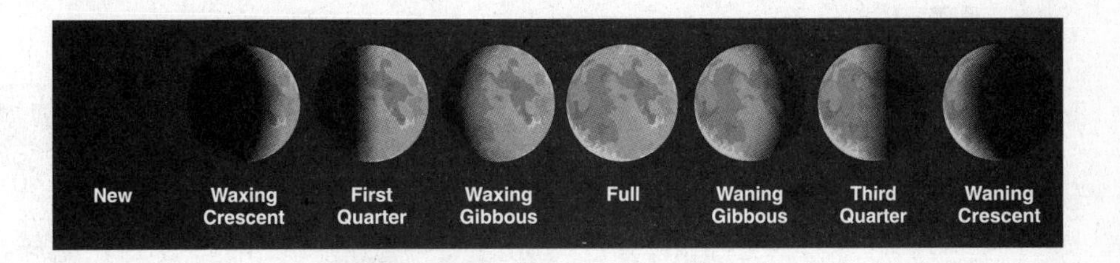

The figure on page 172 shows the positions of the Earth, Moon, and Sun, and how the moon appears from Earth during a period of about 29 days. When the Moon is between the Sun and the Earth, it is called a New Moon. The moon is not visible during this phase because the illuminated side of the moon is facing away from the Earth. As the moon revolves around the Earth, a portion of the illuminated side of the Moon becomes observable from Earth. More and more of the illuminated side of the Moon becomes observable from Earth until it reaches about 14 days into this cycle. At that point, there is a Full Moon, and the Earth, Moon, and Sun are positioned in a way that the Earth is generally between the Moon and Sun. As the Moon's revolution continues, less and less of its illuminated side is observable until finally there is a New Moon, or no visible moon again. These phases of the Moon occur in a regular and predictable manner.

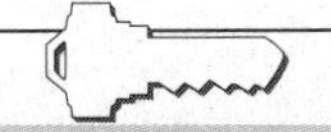

KEY CONCEPT

Only one side of the moon can be observed from Earth because the Moon revolves around the Earth at the same speed that it rotates.

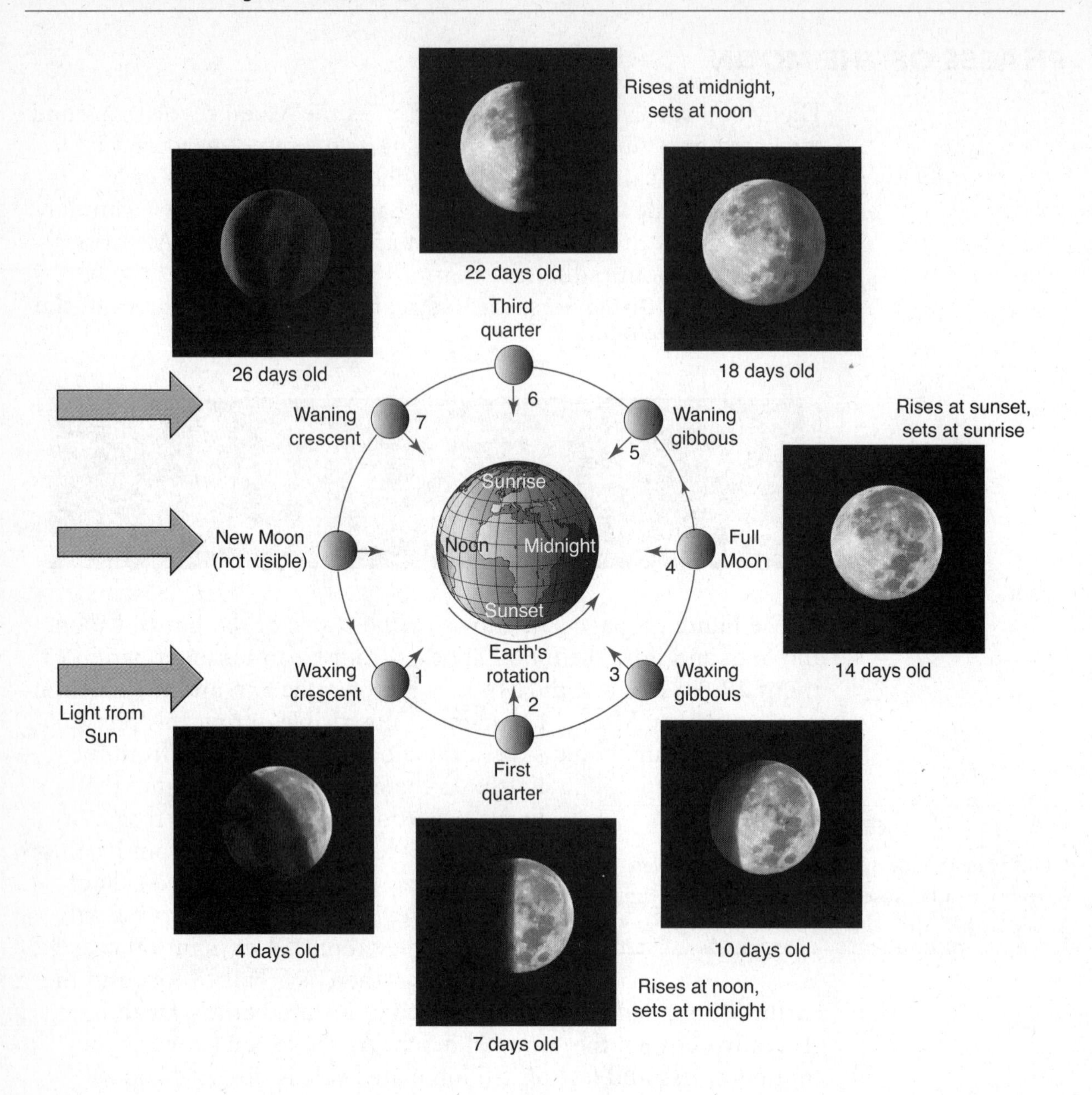

Rises at midnight,
sets at noon
22 days old
26 days old
18 days old
Third
quarter
6
Waning
crescent
7
Waning
gibbous
5
Rises at sunset,
sets at sunrise
Sunrise
New Moon
(not visible)
Noon
Midnight
Full
Moon
4
14 days old
Sunset
Earth's
rotation
Waxing
crescent
1
3
Waxing
gibbous
2
Light from
Sun
First
quarter
4 days old
10 days old
Rises at noon,
sets at midnight
7 days old

Example 13

Which of these is the next moon phase of the sequence?

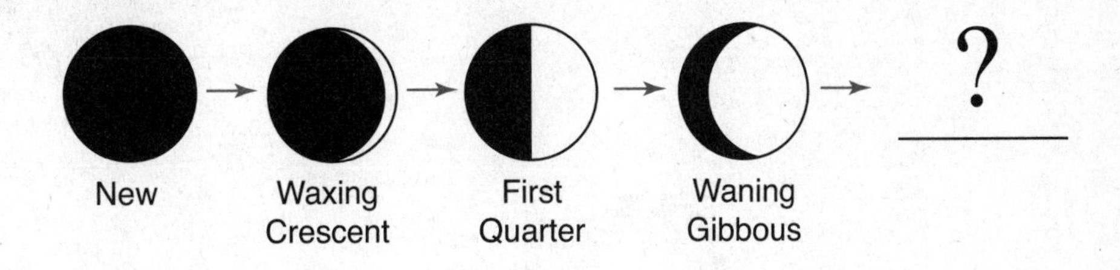

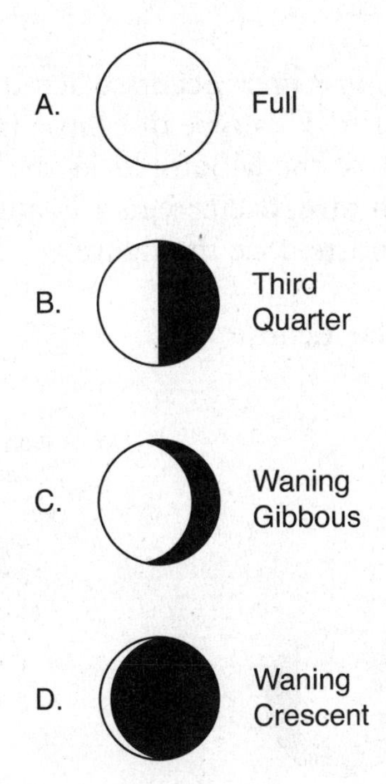

See page 182 for the answer.

ECLIPSES

When the Earth, Moon, and sun are exactly aligned, an eclipse takes place. There are two general types of eclipses in the Earth-Moon-Sun system, lunar eclipses and solar eclipses. A lunar eclipse occurs when the Moon is full and the Earth is between the Sun and the Moon. This type of eclipse lasts for about an hour and begins when the Earth blocks the sunlight from the illuminated surface of the Moon. (See the figure at the top of page 174.)

Lunar Eclipse

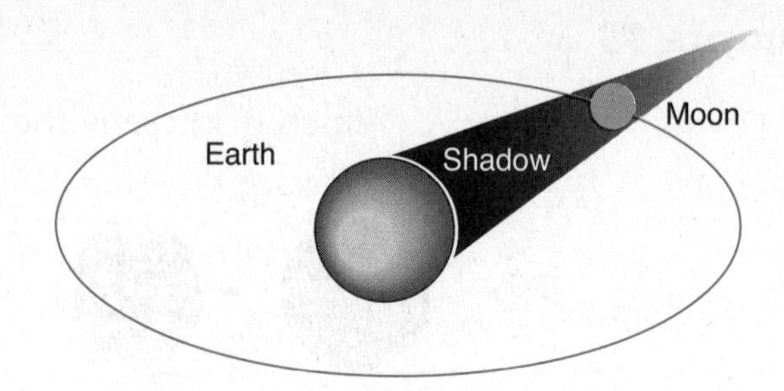

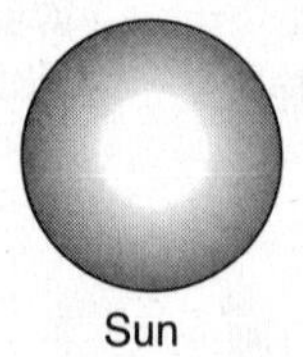

A solar eclipse occurs when the Moon is directly between the Sun and Earth. This type of eclipse lasts for only a few minutes as the shadow of the Moon blocks the light from the sun. Eclipses, through rare, occur regularly and their occurrence can be predicted by scientists. (See the figure.)

Solar Eclipse

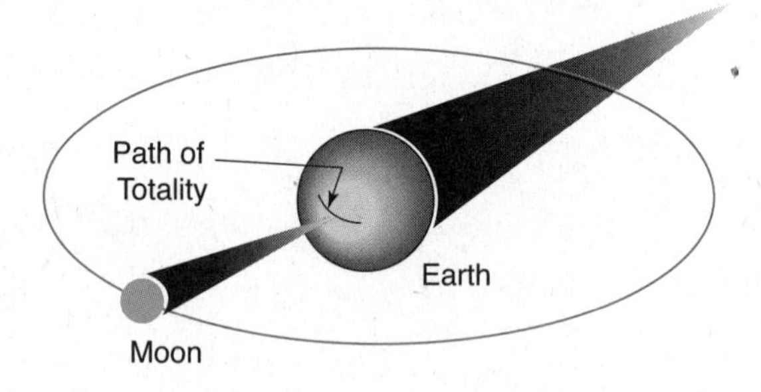

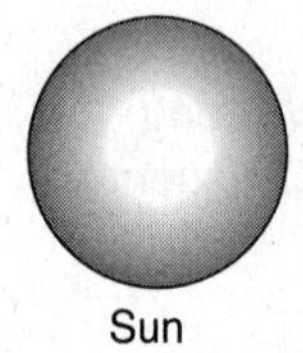

TIDES

The motion of the Earth, Moon, and Sun influences the Earth's oceans. Tides are motions of oceans' water caused by the gravitational attraction between Earth, Moon and Sun. Just as with the phases of the moon and eclipses, tides occur regularly and can be predicted. Two times each day the ocean waters rise and fall mainly due to the gravitational force of the Moon.

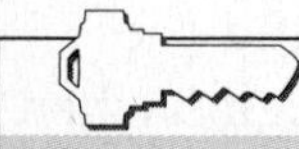

KEY CONCEPT

Eclipses occur when the Sun, Moon, and Earth are aligned.

The Sun's gravitational force also pulls on the ocean water. Twice per month, when the Earth, Moon, and Sun are lined up, the highest tides occur; these are called spring tides and occur during new and full moons. Twice per month when the Sun and Moon are at a 90-degree angle relative to the Earth, the tides are at their minimum level; these are called neap tides. (See the figure.)

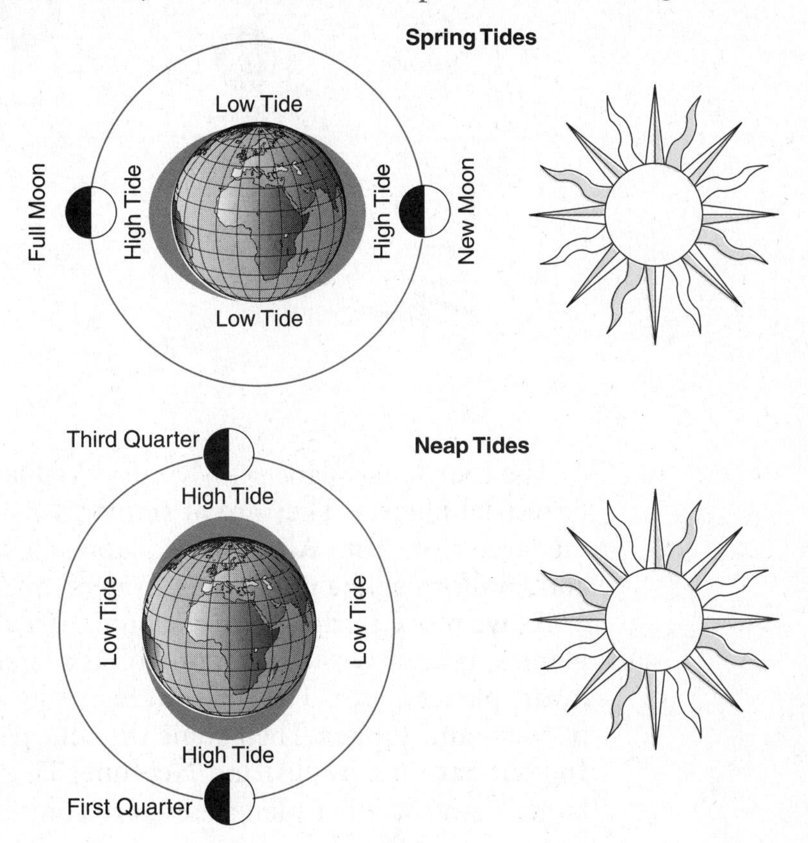

PLANETS IN OUR SOLAR SYSTEM

The planets and other bodies in our solar system can be observed by viewing our night sky. Telescopes improve our ability to observe our solar system by magnifying our view.

Our solar system consists of the Sun and the celestial bodies, such as planets, dwarf planets, asteroids, and meteoroids, revolving around it. The planets, in increasing distance from the Sun, are Mercury, Venus, Earth, Mars, Jupiter, Saturn, Uranus, and Neptune. The dwarf planets in our solar system are Ceres, Pluto, and Eris. The planets orbit around the Sun due to the gravitational force of the Sun. Our solar system is part of the Milky Way galaxy, which is just one of the billion galaxies that make up the universe. (See the figure on page 175.)

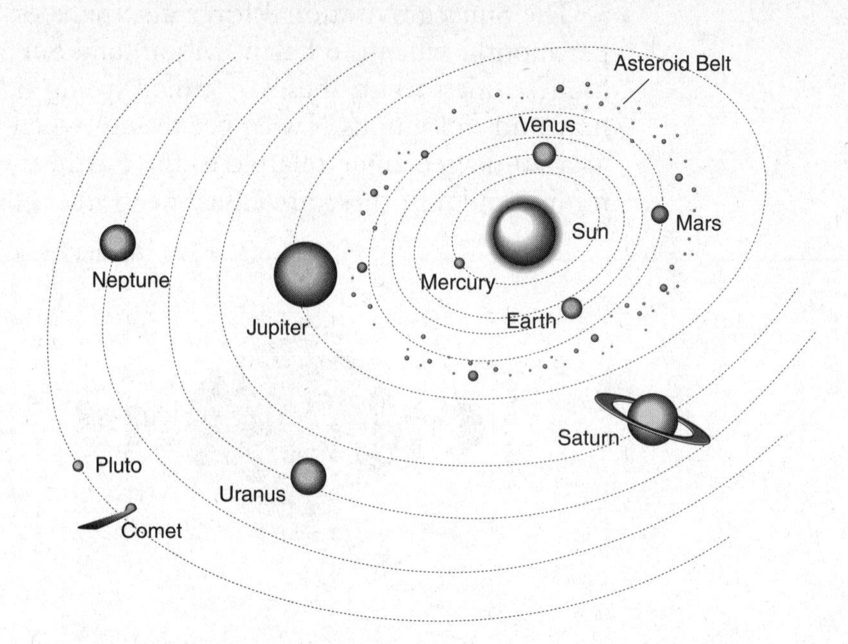

The four inner planets—**Mercury**, **Venus**, **Earth**, and **Mars**—are terrestrial planets. They are of similar density and size, but Earth is the largest of them. All of these planets have a solid surface of rock and landforms, like mountains, craters, and valleys.

As we move farther from the Sun, after the four terrestrial planets, is an **asteroid belt**, composed of many asteroids and the dwarf planet Ceres. The asteroid belt is located between the orbits of Mars and Jupiter. Then come the four gas giant outer planets—**Jupiter**, **Saturn**, **Uranus**, and **Neptune**. These planets are much larger than the inner planets and are composed mainly of gases rather than solid rock. These outer planets also have elaborate ring systems and many moons. Jupiter has about 63 moons! Past the orbit of Neptune is the **Kuiper belt**, which is composed of many icy bodies and the dwarf planet Pluto.

Example 14

Which of the planets below is a terrestrial planet with no atmosphere?

A. Jupiter

B. Uranus

C. Mercury

D. Saturn

See page 182 for the answer.

Comets can also be found in our solar system. Comets are small icy objects that travel around the Sun in highly elliptical orbits. Comets, mostly originating in the Kuiper belt and other far-off regions, produce a glowing head and long tail. The head, or nucleus, of the comet is often described as a "dirty snowball" as it is made of dust, ice, and other chemical compounds. The long tail of a comet always points away from the Sun.

THE SUN AND OTHER STARS

At the center of our solar system is the Sun. The Sun is the major source of energy for phenomena on the Earth's surface, such as growth of plants, winds, and the water cycle. The Sun is a star; because it is much closer to the Earth than any other star, it appears bright and large in the sky, while other stars are tiny dots of light. The Sun is about 93 million miles away from the Earth, yet it is our main source of energy, and we are dependent upon it.

As a star, the Sun shares characteristics with other starts. Stars are large bodies of gas that give off light and energy generated by nuclear reactions. Stars vary in size, brightness, and color. The Sun, although large in comparison to the Earth, is only a middle-sized star. The color of a star is related to the temperature of its surface. Blue-white stars are much hotter than yellow stars (like the Sun), and yellow stars are hotter than red stars.

A star goes through a life cycle similar to organisms; they are born, and they die out, yet their matter is not destroyed. A star's life cycle is predictable based on its initial mass. When some stars begin to die, they become large red stars called red giants. Some stars may explode during a **supernova**. Very massive stars will collapse due to their own gravitational force and turn into a black hole.

SUMMARY

In this chapter, we explored Earth and space science. We looked at weather and climate, soil, minerals, and rocks, changes in the Earth's system, and the Earth in space.

Weather is the state of the atmosphere at a given place and time. Weather conditions are always changing and include temperature, moisture, winds, air pressure, and visibility. Climate is the general weather conditions in a given area over a long period of time. It is affected mostly by latitude, elevation, and location relative to large bodies of water and mountain ranges. The circulation, or movement of water through an ecosystem, is called the water cycle; it is composed of the processes of evaporation, condensation, precipitation, and runoff.

The atmosphere is a mixture of gases surrounding the Earth and consists mostly of nitrogen and oxygen gases. The air is much denser closer to the surface of the Earth. The Earth's atmosphere is heated unevenly. Some areas, like the equator, receive more direct sunlight, while other areas, like the poles, receive sunlight from a less direct, lower angle. This uneven heating of the atmosphere causes winds. Winds blow from areas of high air pressure to regions of low air pressure. The curving of the path of global winds is called the Coriolis effect and is due to Earth's rotation. These global patterns of air circulation even influence weather on a local scale.

Thousands of scientists work around the globe to collect data to make weather predictions. However, weather forecasting is not an exact science. Meteorologists plot atmospheric data, like air pressure, on weather maps. A weather map can be used to represent weather conditions in a given area and predict the temperature and precipitation for several days.

The solid part of the Earth mostly consists of rocks and soil. Soil consists of weathered rocks and decomposed organic material from dead plants, animals, and bacteria. Many factors influence the formation of soil, such as the bedrock, climate, topography and slope of the area, organisms that live in the area, and most importantly, time.

Rocks are made of one or more minerals. Minerals are naturally occurring, inorganic substances that have a definite chemical composition. The physical properties such as color, hardness, luster, and streak can be used to help identify minerals. Rocks are classified into three basic groups based on how the rock formed: igneous, sedimentary, or metamorphic. Rocks are formed, changed, melted, and reformed through the rock cycle, which is driven by the water cycle and plate tectonics.

Earth's interior consists of the lithosphere, the asthenosphere, the mantle, and a dense, metallic core. Plate tectonics is the theory that the lithosphere is broken up into plates that slowly float over the asthenosphere, causing earthquakes, volcanic eruptions, and mountain building. The motion of the plates is due to convection currents in the asthenosphere. Earthquakes often occur at the boundaries between lithospheric plates when there is movement along a fault and the accumulated pressure is released as energy. During an earthquake, the energy is released in the form of seismic waves.

Plate tectonics is a constructive force, building new landforms along plate boundaries. At convergent plate boundaries, plates collide; at divergent plate boundaries, plates move apart; and, at transform plate boundaries, plates slide past each other. The interactions between plates at these boundaries are affected by the density of the rocks in the plates. Various landforms are associated with each type of plate boundary.

Weathering and erosion wear down the Earth's surface. Physical weathering, or mechanical weathering, is the disintegration of rocks into smaller particles. Chemical weathering usually involves a breakdown, or change, in the chemical composition of rocks. After rocks are weathered, erosion by wind, water, and ice may move the soil and rocks to new locations.

The Earth processes we see today, including weathering, erosion, and movement of lithospheric plates, are similar to those that occurred in the past. During Earth's history, occasional catastrophes, such as the impact of an asteroid or comet, occurred. Sedimentary rocks and fossils provide important evidence of how life and environmental conditions have changed throughout Earth's history.

Time is related to motion within the Earth-Moon-Sun system. One day on our planet is the length of time for the Earth to complete one rotation. One month is roughly equivalent to the length of time for the Moon to orbit around the Earth. One year is the length of time for the Earth to complete one revolution around the sun. Motions within the Earth-Moon-Sun system affect seasons, tides, phases of the moon, and eclipses.

Only one side of the moon can be observed from Earth because the Moon revolves around the Earth at the same speed that it rotates. When the Earth, Moon, and Sun are exactly aligned, an eclipse takes place. A lunar eclipse occurs when the Moon is full and the Earth is between the Sun and the Moon. A solar eclipse occurs when the Moon is directly between the Sun and Earth. Tides are motions of oceans' water caused by the gravitational attraction between Earth, Moon, and Sun. Spring tides (highest tidal range) and neap tides (smallest tidal range) occur twice per month.

Our solar system consists of the Sun and the planets, dwarf planets, asteroids, and meteoroids revolving around it. The inner planets are quite different from the outer planets. Comets are found in our solar system and are small icy objects that travel around the Sun in highly elliptical orbits. As a star, the Sun shares characteristics with other stars.

KEY TERMS FROM THIS CHAPTER

water cycle
evaporation
transpiration
condensation
precipitation
runoff
atmosphere
thermometer
sea breeze
land breeze
global winds
Coriolis effect
ocean currents
weather
meteorologists
barometer
air mass
front
cold front
warm front
weather map
climate
crust
lithosphere
mantle
asthenosphere
outer core
inner core
plate tectonics
continental drift
sea floor spreading
earthquakes
fault
seismic waves
seismographs
Richter scale
convergent plate boundaries
divergent plate boundaries
transform plate boundaries
subduct
weathering
physical weathering
mechanical weathering
chemical weathering
erosion
fossils
law of superposition
law of lateral continuity
soil
humans
horizons
rock
mineral
hardness
Moh's hardness scale
luster
streak
magma
igneous rocks
lava
basalt
granite
sediments
lithification
sedimentary rocks
metamorphic rocks
metamorphism
geology
rock cycle
comets
supernova
rotation
revolution
phases of the Moon
eclipse
lunar eclipse
solar eclipse
tides
spring tides
neap tides
telescope
solar system

Mercury	Jupiter
Venus	Saturn
Earth	Uranus
Mars	Neptune
asteroid belt	Kuiper belt

ANSWERS TO EXAMPLES

1. **A** Solar energy (energy from the Sun) drives the water cycle causing the evaporation of water.
2. **C** There is about 78% nitrogen gas, 20% oxygen gas, and 0.93% argon in the atmosphere. (Carbon dioxide makes up about 0.038% of the gas in the atmosphere.)
3. **D** The currents in the Atlantic Ocean off the coast of New Jersey travel north then east toward northern Europe.
4. **B** City B had low temperatures needed for snow to form, nimbostratus clouds characteristic of precipitation, and a falling barometer reading that could indicate a low-pressure system arriving. Low-pressure systems can sometimes bring precipitation.
5. **B** The humus in soils is rich in organic material and nutrients.
6. **D** Hard body parts are more likely to be preserved as fossils. Soft body parts, like organs, tend to decompose before being preserved in rock. On very rare circumstances, soft body material can be preserved.
7. **A** Weathering and erosion cause the breakdown of rock into sediments, not into igneous rocks.
8. **C** The outer core is believed to be made of liquid metal. Even though the mantle behaves like a fluid as it slowly flows in convection currents, it is composed mainly of solid rock.
9. **C** Lithospheric plates can contain both continental crust and oceanic crust, and they move at a rate of a few centimeters per year.
10. **A** Volcanoes form at plate boundaries. The other three choices are usually formed during the erosion of the surface mostly by water.

11. **A** It takes the Earth one year, or 365 days, to completely revolve around the Sun.

12. **C** It is a common misconception that seasons occur because the Earth is closer to the Sun in the summer time. This is wrong. The distance between the Earth and Sun does NOT influence seasonal changes. Seasons are caused by variations in the amount of the Sun's energy hitting the surface, which is due to the tilt of the Earth and the length of the day.

13. **A** The next moon phase is a Full Moon.

14. **C** Mercury is a terrestrial planet and lacks an atmosphere. The other planets—Jupiter, Uranus, and Saturn—are gas giants.

PRACTICE QUIZ

For each of the questions or incomplete statements, choose the best of the answer choices given. Turn to page 242 for the answers.

1. When hiking up the Earth's tallest peaks, why do some climbers bring oxygen equipment to help them breathe?

   A. The denser air has less oxygen at high elevations.
   B. The less dense air has less oxygen at high elevations.
   C. There is no air at the top of mountains.
   D. They use the oxygen to replenish the atmosphere.

2. When limestone is subjected to heat and pressure, marble forms. What type of rock is marble?

   A. sedimentary
   B. igneous
   C. metamorphic
   D. magma

3. Which of the models below shows the correct relationship between the Earth-Moon-Sun system?

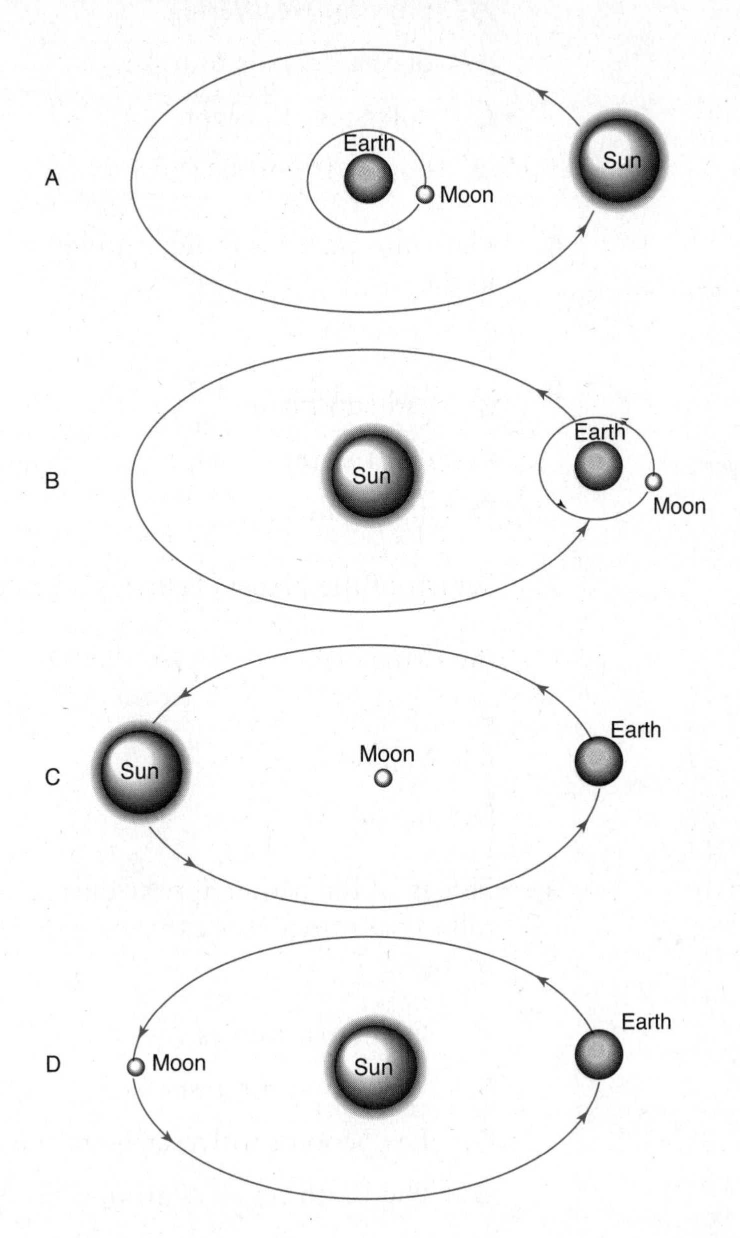

4. The seismic waves of energy released during earthquakes are detected and recorded by a

A. telescope.
B. barometer.
C. seismograph.
D. thermometer.

5. Water trapped within the cracks of rocks may freeze and cause

   A. physical weathering.
   B. quartz crystals to form.
   C. volcanoes to erupt.
   D. humus to form.

6. Plate movement is thought to be caused by convection currents in the

   A. crust.
   B. asthenosphere.
   C. outer core.
   D. atmosphere.

7. Which of the planets below is a gas giant?

   A. Mercury
   B. Uranus
   C. Venus
   D. Earth

8. Phases of the Moon appear during each month. Which of the following statements explains why the Moon's appearance changes?

   A. The Earth rotates.
   B. The Moon rotates.
   C. The Moon revolves around the Earth.
   D. The Earth revolves around the Moon.

9. The picture shows the positions of the Earth, Moon, and Sun during a lunar eclipse. A lunar eclipse occurs

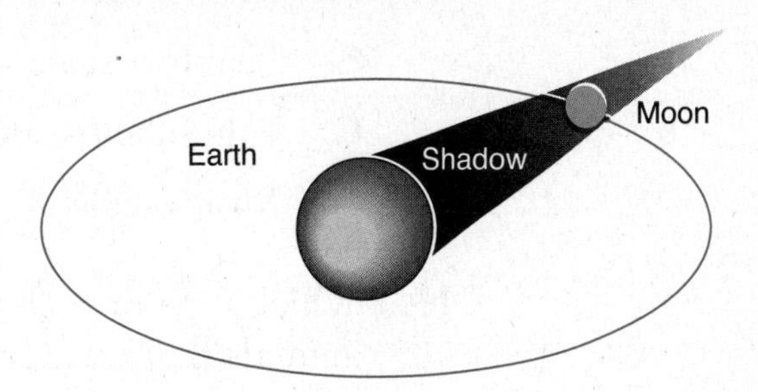

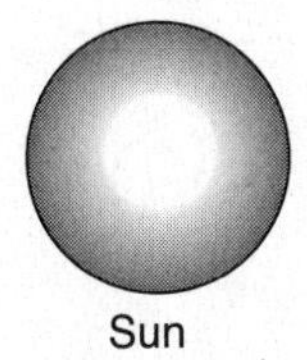

A. during the New Moon phase.

B. when the moon is between the Sun and Earth.

C. when the Sun is between the Earth and Moon.

D. when the Earth is between the Moon and Sun.

10. During which of the dates below does New Jersey experience its longest day of the year?

A. March 20th

B. June 21st

C. September 22nd

D. December 21st

11. Which instrument assists scientists with their observations of celestial objects?

A. telescope

B. microscope

C. barometer

D. thermometer

12. What scale is used to determine the hardness of minerals?

A. Richter scale
B. Beaufort scale
C. Fahrenheit scale
D. Moh's scale

13. Which type of rock will most likely contain the fossils of animals that once lived in the ocean?

A. igneous rock
B. sedimentary rock
C. metamorphic rock
D. basalt

14. Which layer of the Earth is the hottest?

A. crust
B. mantle
C. core
D. asthenosphere

15. Which of these is the next moon phase of the sequence?

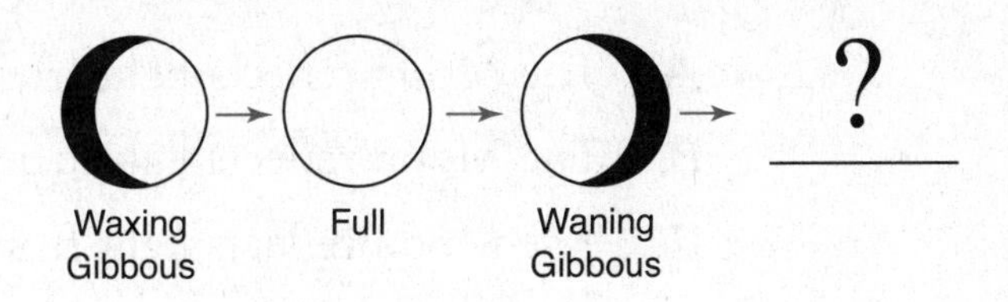

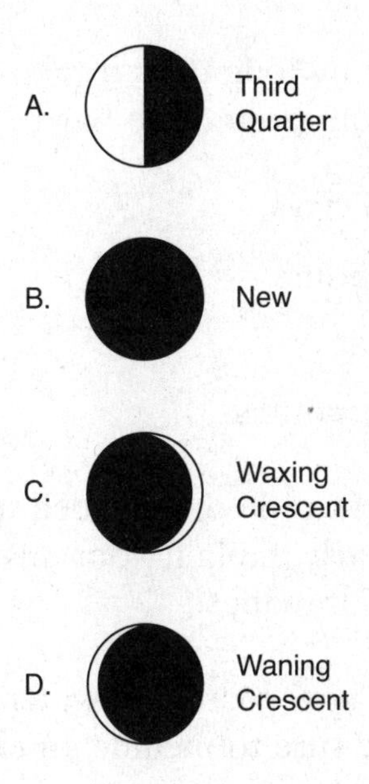

16. The Sun is a

A. planet.
B. star.
C. galaxy.
D. dwarf planet.

17. What are two common causes of physical weathering?

A. wind and water
B. wind and chemical reactions
C. chemical reactions and acid rain
D. acid rain and wind

18. How can we see the Moon since it produces no light of its own?

A. The Moon reflects light from the Sun.

B. The Moon reflects light from other stars.

C. The Moon reflects light from the Earth.

D. The craters of the Moon capture light and release it at night.

19. About how long does it take the Moon to complete one revolution around the Earth?

A. 365 days

B. 24 hours

C. 29 days

D. 60 minutes

Respond fully to the open-ended question that follows. Show your work and clearly explain your answer. You may use words, tables, diagrams, or drawings.

20. Explain how a drop of water can travel through the water cycle. Be sure to include an explanation of all of the processes of the water cycle.

PRACTICE TEST 1 ANSWER SHEET

Part 1

1. Ⓐ Ⓑ Ⓒ Ⓓ	6. Ⓐ Ⓑ Ⓒ Ⓓ	11. Ⓐ Ⓑ Ⓒ Ⓓ
2. Ⓐ Ⓑ Ⓒ Ⓓ	7. Ⓐ Ⓑ Ⓒ Ⓓ	12. Ⓐ Ⓑ Ⓒ Ⓓ
3. Ⓐ Ⓑ Ⓒ Ⓓ	8. Ⓐ Ⓑ Ⓒ Ⓓ	13. Ⓐ Ⓑ Ⓒ Ⓓ
4. Ⓐ Ⓑ Ⓒ Ⓓ	9. Ⓐ Ⓑ Ⓒ Ⓓ	14. Ⓐ Ⓑ Ⓒ Ⓓ
5. Ⓐ Ⓑ Ⓒ Ⓓ	10. Ⓐ Ⓑ Ⓒ Ⓓ	15. Ⓐ Ⓑ Ⓒ Ⓓ

16.

Part 2

17. Ⓐ Ⓑ Ⓒ Ⓓ
18. Ⓐ Ⓑ Ⓒ Ⓓ
19. Ⓐ Ⓑ Ⓒ Ⓓ
20. Ⓐ Ⓑ Ⓒ Ⓓ
21. Ⓐ Ⓑ Ⓒ Ⓓ
22. Ⓐ Ⓑ Ⓒ Ⓓ
23. Ⓐ Ⓑ Ⓒ Ⓓ
24. Ⓐ Ⓑ Ⓒ Ⓓ
25. Ⓐ Ⓑ Ⓒ Ⓓ
26. Ⓐ Ⓑ Ⓒ Ⓓ
27. Ⓐ Ⓑ Ⓒ Ⓓ
28. Ⓐ Ⓑ Ⓒ Ⓓ
29. Ⓐ Ⓑ Ⓒ Ⓓ
30. Ⓐ Ⓑ Ⓒ Ⓓ
31. Ⓐ Ⓑ Ⓒ Ⓓ

32.

Part 3

33. Ⓐ Ⓑ Ⓒ Ⓓ
34. Ⓐ Ⓑ Ⓒ Ⓓ
35. Ⓐ Ⓑ Ⓒ Ⓓ
36. Ⓐ Ⓑ Ⓒ Ⓓ
37. Ⓐ Ⓑ Ⓒ Ⓓ
38. Ⓐ Ⓑ Ⓒ Ⓓ
39. Ⓐ Ⓑ Ⓒ Ⓓ
40. Ⓐ Ⓑ Ⓒ Ⓓ
41. Ⓐ Ⓑ Ⓒ Ⓓ
42. Ⓐ Ⓑ Ⓒ Ⓓ
43. Ⓐ Ⓑ Ⓒ Ⓓ
44. Ⓐ Ⓑ Ⓒ Ⓓ
45. Ⓐ Ⓑ Ⓒ Ⓓ
46. Ⓐ Ⓑ Ⓒ Ⓓ
47. Ⓐ Ⓑ Ⓒ Ⓓ

48.

Part 4

49 Ⓐ Ⓑ Ⓒ Ⓓ	54. Ⓐ Ⓑ Ⓒ Ⓓ	59. Ⓐ Ⓑ Ⓒ Ⓓ
50. Ⓐ Ⓑ Ⓒ Ⓓ	55. Ⓐ Ⓑ Ⓒ Ⓓ	60. Ⓐ Ⓑ Ⓒ Ⓓ
51. Ⓐ Ⓑ Ⓒ Ⓓ	56. Ⓐ Ⓑ Ⓒ Ⓓ	61. Ⓐ Ⓑ Ⓒ Ⓓ
52. Ⓐ Ⓑ Ⓒ Ⓓ	57. Ⓐ Ⓑ Ⓒ Ⓓ	62. Ⓐ Ⓑ Ⓒ Ⓓ
53. Ⓐ Ⓑ Ⓒ Ⓓ	58. Ⓐ Ⓑ Ⓒ Ⓓ	63. Ⓐ Ⓑ Ⓒ Ⓓ

64.

PRACTICE TEST 1

PART 1

For each of the questions or incomplete statements below, choose the best of the answer choices given.

1. A mosquito biting a person is an example of which type of relationship?

   A. parasite/host
   B. consumer/producer
   C. predator/prey
   D. consumer/decomposer

2. Which of the following gases is a compound?

   A. carbon dioxide (CO_2)
   B. oxygen (O_2)
   C. ozone (O_3)
   D. helium (He)

3. Why is there night and day on Earth?

   A. The sun rotates.
   B. The Earth rotates.
   C. The Earth's axis is tilted.
   D. The Earth revolves around the Sun.

4. Plate movement is thought to be caused by convection currents in the

   A. crust.
   B. asthenosphere.
   C. outer core.
   D. atmosphere.

5. When light strikes some objects, the light bounces off of it. These light rays are

   A. radiated.
   B. refracted.
   C. transmitted.
   D. scattered.

6. During the evening, the land cools faster than the seawater. The air over sea is warmer and has a lower air pressure than the air over the land. The cooler air from the land side blows in the direction of the water. What is the name of this type of wind?

   A. global
   B. land breeze
   C. sea breeze
   D. solar wind

7. A green insect lives on a tree among many leaves. How does the insect's green color help it survive?

   A. It helps the insect locate mates.
   B. It allows the insect to make its own food through photosynthesis.
   C. It keeps the insect warm.
   D. It makes the insect hard to see when it sits among leaves.

8. Using the Periodic Table, which of the following elements has 12 protons?

1	2	3	4	5	6	7	8	9	10	11	12	13	14	15	16	17	18
2 **H** 1.01																	2 **He** 4.00
3 **Li** 6.94	4 **Be** 9.01											5 **B** 10.81	6 **C** 12.01	7 **N** 14.01	8 **O** 16.00	9 **F** 19.00	10 **Ne** 20.18
11 **Na** 22.99	12 **Mg** 24.30											13 **Al** 26.98	14 **Si** 28.09	15 **P** 30.97	16 **S** 32.07	17 **Cl** 35.45	18 **Ar** 39.95
19 **K** 30.10	20 **Ca** 40.08	21 **Sc** 44.96	22 **Ti** 47.88	23 **V** 50.94	24 **Cr** 52.00	25 **Mn** 54.94	26 **Fe** 55.85	27 **Co** 58.93	28 **Ni** 58.69	29 **Cu** 63.55	30 **Zn** 65.39	31 **Ga** 69.72	32 **Ge** 72.61	33 **As** 74.92	34 **Se** 78.96	35 **Br** 79.90	36 **Kr** 83.80
37 **Rb** 85.47	38 **Sr** 87.62	39 **Y** 88.91	40 **Zr** 91.22	41 **Nb** 92.91	42 **Mo** 95.94	43 **Tc** (97.91)	44 **Ru** 101.07	45 **Rh** 102.91	46 **Pd** 106.42	47 **Ag** 107.87	48 **Cd** 112.41	49 **In** 114.82	50 **Sn** 118.71	51 **Sb** 121.75	52 **Te** 127.60	53 **I** 126.90	54 **Xe** 131.29
55 **Cs** 132.91	56 **Ba** 137.33	57 **La** 138.91	72 **Hf** 178.49	73 **Ta** 180.95	74 **W** 183.85	75 **Re** 186.21	76 **Os** 190.23	77 **Ir** 192.22	78 **Pt** 195.08	79 **Au** 196.97	80 **Hg** 200.59	81 **Tl** 204.38	82 **Pb** 207.2	83 **Bi** 208.98	84 **Po** (208.98)	85 **At** (209.99)	86 **Rn** (222.02)
87 **Fr** (223.02)	88 **Ra** (226.03)	89 **Ac** (227.03)	104 **Rf** (261.11)	105 **Ha** (262.11)	106 **Sg** (263.12)												

58 **Ce** 140.12	59 **Pr** 140.91	60 **Nd** 144.24	61 **Pm** (144.91)	62 **Sm** 150.36	63 **Eu** 151.97	64 **Gd** 157.25	65 **Tb** 158.93	66 **Dy** 162.50	67 **Ho** 164.93	68 **Er** 167.26	69 **Tm** 168.93	70 **Yb** 173.04	71 **Lu** 174.97
90 **Th** 232.04	91 **Pa** 231.04	92 **U** 238.03	93 **Np** (237.05)	94 **Pu** (244.06)	95 **Am** (243.06)	96 **Cm** (247.07)	97 **Bk** (247.07)	98 **Cf** (251.08)	99 **Es** (252.08)	100 **Fm** (257.10)	101 **Md** (258.10)	102 **No** (259.10)	103 **Lr** (262.11)

A. hydrogen

B. magnesium

C. sodium

D. lithium

9. About how long does it take the Moon to complete one revolution around the Earth?

A. 365 days

B. 24 hours

C. 29 days

D. 60 minutes

10. Which of the following organisms in a pond ecosystem makes its own food through photosynthesis?

A. frogs

B. minnows

C. water lily

D. toads

11. Atoms are composed of protons, neutrons, and electrons. What is the electric charge of the electron?

A. positive

B. negative

C. neutral

D. double-charged

12. How can we see the Moon since it produces no light of its own?

A. The Moon reflects light from the Sun.

B. The Moon reflects light from other stars.

C. The Moon reflects light from the Earth.

D. The craters of the Moon capture light and release it at night.

13. Which of the following is in order from simplest to most complex?

A. organs, cells, tissues, organ systems

B. cells, tissues, organs, systems

C. systems, tissues, cells, organs

D. tissues, cells, systems, organs

14. When sunlight shines on Jack's shirt, the shirt appears yellow. Why does the shirt look yellow?

A. The sunlight is yellow and makes all objects appear yellow.

B. The shirt absorbs white light and changes it into yellow.

C. It scatters the yellow part of the light spectrum and absorbs the other light.

D. It absorbs the yellow light from the Sun and scatters all of the other colors.

15. In a forest ecosystem, which of the organisms below break down dead plants and animals?

A. oak trees

B. owls

C. squirrels

D. bacteria

Respond fully to the open-ended question that follows. Show your work and clearly explain your answer. You may use words, tables, diagrams, or drawings.

---

16. We are dependent upon electricity to power our many appliances. Select a source of energy (Hint: any fossil fuel or alternative source of energy), and explain the equipment needed and the energy transformations that must take place to convert the energy into electricity.

PART 2

For each of the questions or incomplete statements below, choose the best of the answer choices given.

17. Which type of rock will most likely contain the fossils of animals that once lived in a stream?

A. igneous rock
B. sedimentary rock
C. metamorphic rock
D. basalt

18. Of the following materials, which is translucent?

A. frosted glass
B. piece of thick cardboard
C. wood door
D. clear glass

19. In which two organs does the digestion of food primarily take place?

A. stomach and small intestine
B. liver and kidneys
C. large intestine and kidneys
D. liver and small intestine

20. Which landform is most likely created by weathering and erosion?

A. volcano
B. midocean ridges
C. mountain ranges
D. canyons

21. Soup mix is made of a variety of ingredients. The figure shows some of the ingredients. Which term best describes soup mix?

Nutrition Facts
Ingredients: dehydrated onions, garlic, dehydrated celery, carrots, salt, spices

A. element

B. atoms

C. compound

D. mixture

22. A class took a field trip to a pond and collected water samples. Using a microscope, students observed green algae in the pond water. What is the role of green algae in the pond ecosystem?

A. producer

B. primary consumer

C. secondary consumer

D. decomposer

23. During which of the dates below does New Jersey experience its shortest day of the year?

A. March 20th

B. June 21st

C. September 22nd

D. December 21st

24. A team of students examined an object. Which of the following is NOT an observation about the object?

A. It is smooth and shiny.

B. It has a mass of 552 grams.

C. It is 21 centimeters long.

D. I think it is metal and used in a machine.

25. When a person touches a hot stove, what carries the message from the fingertips to the brain?

A. veins

B. glands

C. nerves

D. skin cells

26. Which of the following processes occurs in the formation of sedimentary rocks?

A. lithification

B. melting

C. crystallization

D. cooling

27. Traits are transferred from parent to offspring through

A. the sperm and egg.

B. blood.

C. food.

D. organs.

28. Which of the following causes the Sun to appear to set?

A. the rotation of the Moon

B. the rotation of the Sun

C. the rotation of the Earth

D. the revolution of the Earth

29. Students worked in teams in a science class to complete the same experiment. After conducting the investigation and collecting data, one team gathered all the data from the other teams. After comparing their data with the other teams' data, what would indicate that their data are valid?

    A. Their team finished before the other teams.
    B. Many of the other teams recorded similar data.
    C. Another class completed the same experiment.
    D. Their team followed all of the rules and procedures.

30. What is the most common cause of earthquakes?

    A. tsunamis
    B. weathering
    C. movements in Earth's crust
    D. erosion

31. Which of the diagrams illustrates the refraction of light as it travels from air to water?

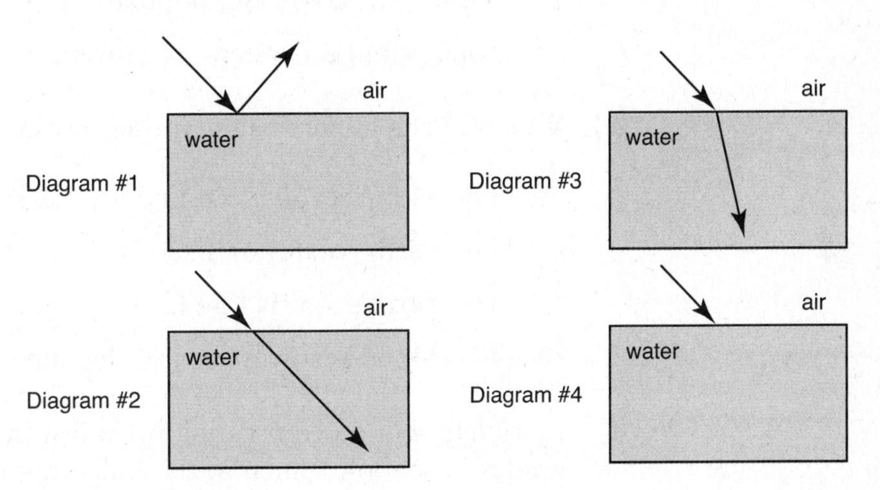

    A. diagram #1
    B. diagram #2
    C. diagram #3
    D. diagram #4

Respond fully to the open-ended question that follows. Show your work and clearly explain your answer. You may use words, tables, diagrams, or drawings.

32. You are given a bucket filled with seawater. Describe the materials needed and the steps that could be used to obtain fresh drinking water from the seawater.

PART 3

For each of the questions or incomplete statements below, choose the best of the answer choices given.

33. Which sequence shows the ecological levels from simplest to most complex?

A. population, community, ecosystem

B. ecosystem, population, community

C. organism, ecosystem, population

D. population, ecosystem, organism

34. Which of the factors below influences seasons on Earth?

A. The Moon revolves around the Earth.

B. The Earth rotates on its axis.

C. The Earth's axis is tilted.

D. The Moon revolves around the Sun.

35. A student wondered if eating breakfast improves students' grades in school. Which of the following best represents a possible hypothesis from this student?

A. Eating breakfast has no affect on attendance at school.

B. If students eat breakfast, the grades will improve in school.

C. If grades improve in school, then students will enjoy breakfast.

D. How does eating breakfast affect a student's grades?

36. Two beakers were placed on a windowsill. One beaker contained 100 mL of water, and the other beaker contained 100 mL of rubbing alcohol. The next day both cups had less liquid in them, but there was less rubbing alcohol than water. What can be concluded from this experiment?

    A. Only some liquids evaporate.
    B. Liquids can only evaporate in the presence of sunlight.
    C. Some liquids evaporate faster than others.
    D. Water evaporates faster than rubbing alcohol.

37. Red blood cells

    A. carry oxygen to all parts of the body.
    B. fight diseases.
    C. transmit messages to the brain.
    D. digest food particles.

38. What is the source of energy that drives the water cycle?

    A. solar energy
    B. global winds
    C. Sun's gravity
    D. Earth's gravity

39. A student wanted to test how temperature affects the sprouting of alfalfa seeds. She set up five dishes and placed ten alfalfa seeds on a moist paper towel in each dish. Each of the five dishes were kept at a different temperature, 10°C, 15°C, 20°C, 25°C, and 30°C. Which of these is the independent, or manipulated, variable in this experiment?

    A. the number of seeds
    B. the number of dishes
    C. the temperature of each dish
    D. the moist paper towel

40. The nucleus of most atoms consists of

A. neutrons and electrons.
B. protons and neutrons.
C. protons and electrons.
D. electrons.

41. Which of the planets below is a gas giant?

A. Mars
B. Neptune
C. Venus
D. Earth

42. The first organism in a food chain must be a

A. producer.
B. herbivore.
C. decomposer.
D. consumer.

43. Which of the following statements is a hypothesis?

A. Why do your ears pop in an airplane?
B. Deer are herbivores.
C. If the temperature of the water increases, then more sugar can dissolve in it.
D. How do earthworms react in moist environments?

44. What is the product of meiosis?

A. new body cells
B. sex cells
C. chromosomes
D. new organisms

45. Which gas is most abundant in our atmosphere?

A. nitrogen

B. carbon dioxide

C. oxygen

D. water vapor

46. What would happen to this food chain if humans sprayed insecticide?

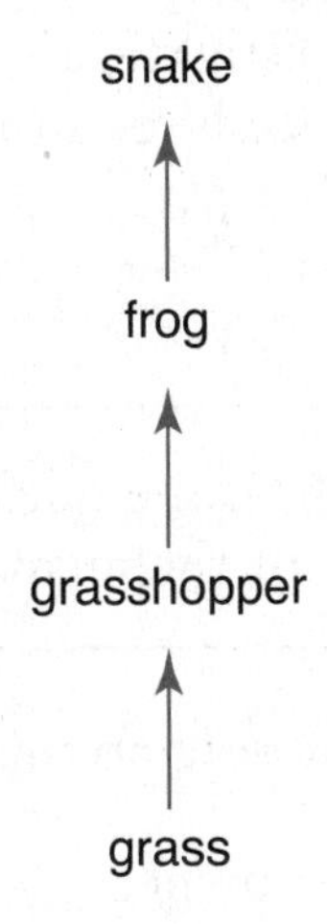

A. The number of snakes would increase.

B. The number of grasshoppers would increase.

C. The number of frogs would decrease.

D. The grass would stop growing.

47. The distance between one crest on a wave and the next crest of the wave is called a

A. trough.

B. amplitude.

C. wavelength.

D. frequency.

Respond fully to the open-ended question that follows. Show your work and clearly explain your answer. You may use words, tables, diagrams, or drawings.

48. While playing soccer in physical education class, a student noticed that her heart rate had increased while playing and didn't return to normal immediately when she stopped playing. She wants to investigate how long it takes for her heart rate to return to normal after playing. How should she design her investigation? What materials does she need and what procedures should she follow?

Part 4

For each of the questions or incomplete statements below, choose the best of the answer choices given.

49. The role an organism plays in an ecosystem is called a

- **A.** niche.
- **B.** habitat.
- **C.** food web.
- **D.** population.

50. A child can inherit traits

- **A.** from both his parents.
- **B.** from either one of his parents.
- **C.** only from his father.
- **D.** only from his mother.

51. Phases of the moon appear during each month. Which of the following explanations explains why the Moon's appearance changes?

A. The Earth rotates.

B. The Moon rotates.

C. The Moon revolves around the Earth.

D. The Earth revolves around the Moon.

52. In the model of an atom below, what is the arrow indicating?

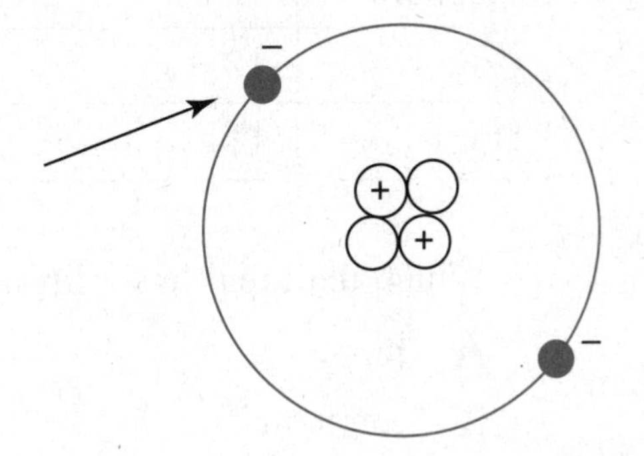

A. proton

B. electron

C. neutron

D. nucleus

53.

LIFE SPAN AND GESTATIONAL PERIOD OF COMMON MAMMALS

Mammal	Average Life Span (years)	Gestational Period (days)
Human	77	266
Horse	27	330
Elephant	70	645
Cow	18	284
Dog	16	61

Which mammal has a lifespan closest to that of a dog?

A. horse

B. elephant

C. cow

D. dog

54. After swimming in a pool, Sally dried herself off using a towel. The wet towel was left in the Sun. Later, Sally observed that the towel had dried. Which process occurred to make this happen?

A. condensation

B. evaporation

C. precipitation

D. melting

55. While conducting a chemistry investigation, a student mixed a white powder into a beaker filled with a clear liquid. As stated in the student's science journal, "After stirring in the white powder, the liquid became cloudy and then turned light blue." This statement is

A. a problem.

B. a hypothesis.

C. an observation.

D. a conclusion.

56. A student wanted to test how sugar water affects the growth of corn plants. She watered all of the corn plants with a mixture of sugar and water. How could she improve her experiment?

A. Water half of the corn plants with pure water.

B. Warm the sugar and water mixture before watering the plants.

C. Cool the sugar and water mixture before watering the plants.

D. Use different types of plants.

57. Fossil fuels were formed from

A. earthquakes.

B. the metamorphism of rocks.

C. water trapped with rock.

D. the remains of organisms.

58. A female chimpanzee has 48 chromosomes in each of her body cells. How many chromosomes are found in each of her egg cells?

A. 48

B. 96

C. 24

D. 12

59. Which of the following best shows particles of a compound?

A
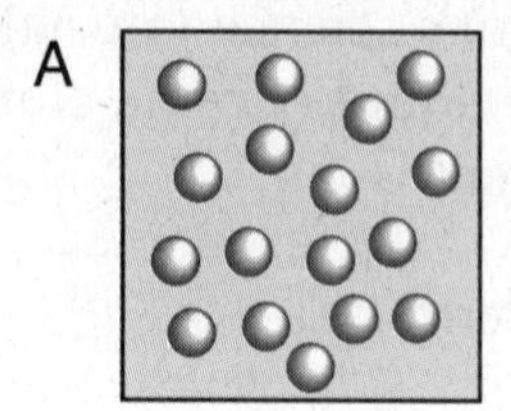

B
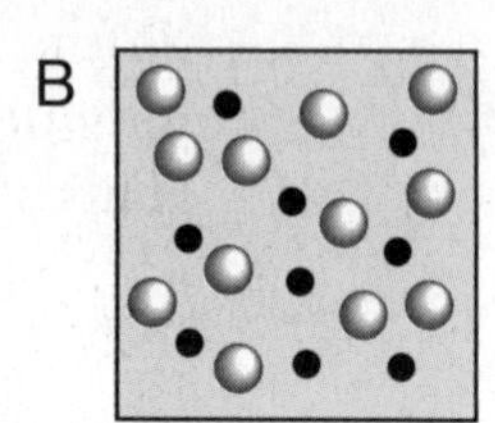

C
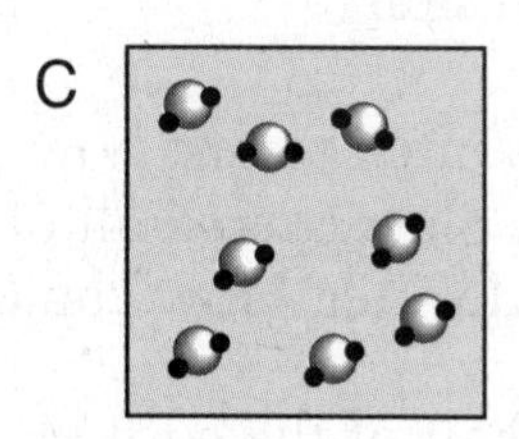

D
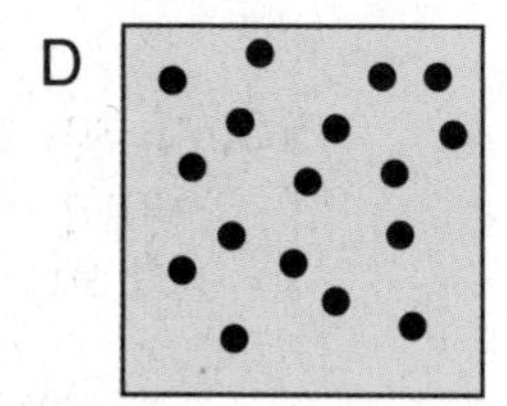

60. How many legs do all insects have?

A. 2

B. 4

C. 6

D. 8

61. Which of the following molecules does not contain two atoms?

A. O_2

B. NaCl

C. N_2

D. H_2O

62. Which is composed of a group of specialized, or similar, cells working together?

A. tissue

B. organ

C. organ system

D. organism

63. Which of the statements below correctly describes one of the water cycle processes?

A. Precipitation can occur when water vapor loses energy and forms liquid water droplets suspended in the air.

B. Condensation can occur when liquid water molecules fall to Earth as snow, rain, or sleet.

C. Evaporation can occur when water gains energy from the Sun and changes into water vapor.

D. Runoff can occur when water vapor gains energy and moves higher in the atmosphere.

Respond fully to the open-ended question that follows. Show your work and clearly explain your answer. You may use words, tables, diagrams, or drawings.

---

64. Doctors use various tools to help them do their job. How do a doctor's tools help him or her do his or her job? How have microscopes helped doctors? How would a doctor's job be different if he did not have microscopes or other tools?

PRACTICE TEST 2 ANSWER SHEET

Part 1

1. Ⓐ Ⓑ Ⓒ Ⓓ
2. Ⓐ Ⓑ Ⓒ Ⓓ
3. Ⓐ Ⓑ Ⓒ Ⓓ
4. Ⓐ Ⓑ Ⓒ Ⓓ
5. Ⓐ Ⓑ Ⓒ Ⓓ
6. Ⓐ Ⓑ Ⓒ Ⓓ
7. Ⓐ Ⓑ Ⓒ Ⓓ
8. Ⓐ Ⓑ Ⓒ Ⓓ
9. Ⓐ Ⓑ Ⓒ Ⓓ
10. Ⓐ Ⓑ Ⓒ Ⓓ
11. Ⓐ Ⓑ Ⓒ Ⓓ
12. Ⓐ Ⓑ Ⓒ Ⓓ
13. Ⓐ Ⓑ Ⓒ Ⓓ
14. Ⓐ Ⓑ Ⓒ Ⓓ
15. Ⓐ Ⓑ Ⓒ Ⓓ
16.

Part 2

17. Ⓐ Ⓑ Ⓒ Ⓓ
18. Ⓐ Ⓑ Ⓒ Ⓓ
19. Ⓐ Ⓑ Ⓒ Ⓓ
20. Ⓐ Ⓑ Ⓒ Ⓓ
21. Ⓐ Ⓑ Ⓒ Ⓓ
22. Ⓐ Ⓑ Ⓒ Ⓓ
23. Ⓐ Ⓑ Ⓒ Ⓓ
24. Ⓐ Ⓑ Ⓒ Ⓓ
25. Ⓐ Ⓑ Ⓒ Ⓓ
26. Ⓐ Ⓑ Ⓒ Ⓓ
27. Ⓐ Ⓑ Ⓒ Ⓓ
28. Ⓐ Ⓑ Ⓒ Ⓓ
29. Ⓐ Ⓑ Ⓒ Ⓓ
30. Ⓐ Ⓑ Ⓒ Ⓓ
31. Ⓐ Ⓑ Ⓒ Ⓓ
32.

Part 3

33. Ⓐ Ⓑ Ⓒ Ⓓ	38. Ⓐ Ⓑ Ⓒ Ⓓ	43. Ⓐ Ⓑ Ⓒ Ⓓ
34. Ⓐ Ⓑ Ⓒ Ⓓ	39. Ⓐ Ⓑ Ⓒ Ⓓ	44. Ⓐ Ⓑ Ⓒ Ⓓ
35. Ⓐ Ⓑ Ⓒ Ⓓ	40. Ⓐ Ⓑ Ⓒ Ⓓ	45. Ⓐ Ⓑ Ⓒ Ⓓ
36. Ⓐ Ⓑ Ⓒ Ⓓ	41. Ⓐ Ⓑ Ⓒ Ⓓ	46. Ⓐ Ⓑ Ⓒ Ⓓ
37. Ⓐ Ⓑ Ⓒ Ⓓ	42. Ⓐ Ⓑ Ⓒ Ⓓ	47. Ⓐ Ⓑ Ⓒ Ⓓ

48.

Part 4

49 Ⓐ Ⓑ Ⓒ Ⓓ

50. Ⓐ Ⓑ Ⓒ Ⓓ

51. Ⓐ Ⓑ Ⓒ Ⓓ

52. Ⓐ Ⓑ Ⓒ Ⓓ

53. Ⓐ Ⓑ Ⓒ Ⓓ

54. Ⓐ Ⓑ Ⓒ Ⓓ

55. Ⓐ Ⓑ Ⓒ Ⓓ

56. Ⓐ Ⓑ Ⓒ Ⓓ

57. Ⓐ Ⓑ Ⓒ Ⓓ

58. Ⓐ Ⓑ Ⓒ Ⓓ

59. Ⓐ Ⓑ Ⓒ Ⓓ

60. Ⓐ Ⓑ Ⓒ Ⓓ

61. Ⓐ Ⓑ Ⓒ Ⓓ

62. Ⓐ Ⓑ Ⓒ Ⓓ

63. Ⓐ Ⓑ Ⓒ Ⓓ

64.

PRACTICE TEST 2

PART 1

For each of the questions or incomplete statements below, choose the best of the answer choices given.

1. What causes the Coriolis effect, or the deflection of global winds?

   A. the rotation of the Earth
   B. the revolution of the Earth around the Sun
   C. the rotation of the Moon
   D. the revolution of the Moon around the Earth

2. How are humans classified in our ecosystems?

   A. producers
   B. consumers
   C. decomposers
   D. parasites

3. A ball is dropped from the roof of a building. As the ball falls its potential energy is converted to

   A. solar energy.
   B. nuclear energy.
   C. kinetic energy.
   D. chemical energy.

4. Which biome contains the most diversity of life?

A. desert
B. tropical rainforest
C. tundra
D. grassland

5. Which of the diagrams below best represents the interior layers of the Earth?

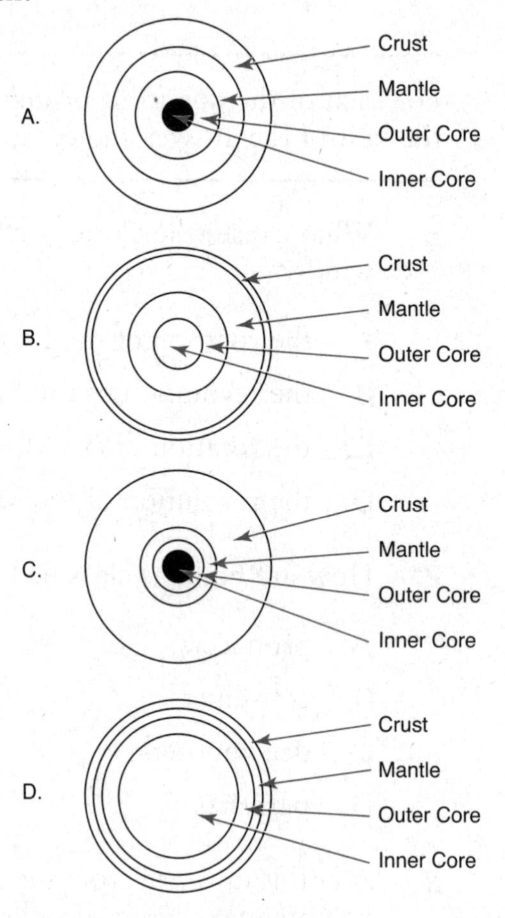

6. Which of the following is an example of a wave?

A. a string vibrating on a guitar
B. a ball being dropped
C. a bird flying in the atmosphere
D. a car traveling down a highway

7. Which cell part allows nutrients and other materials to enter and exit the cell?

A. nucleus

B. cell membrane

C. cytoplasm

D. vacuole

8. Which type of rock is produced by the cooling and solidification of molten rock?

A. igneous rock

B. sedimentary rock

C. metamorphic rock

D. magma

9. Which materials would be most useful if you wanted to separate a mixture of sand and sugar?

A. ruler and balance

B. filter paper and water

C. magnet and a beaker

D. graduated cylinder and a battery

10. Which sequence shows the correct order in a food chain?

A. producers → carnivores → herbivores

B. carnivores → producers → herbivores

C. producers → herbivores → carnivores

D. herbivores → carnivores → producers

11. The rotation of the Earth causes

A. day and night.

B. years.

C. months.

D. seasons.

12. Put the following stages of succession in their proper sequence.

    I. Grasses
    II. Pine trees
    III. Scrub growth
    IV. Lichen/moss

    A. III, IV, II, I
    B. II, I, III, IV
    C. I, II, III, IV
    D. IV, I, III, II

13. The picture shows the positions of the Earth, Moon, and Sun during a solar eclipse. A solar eclipse occurs

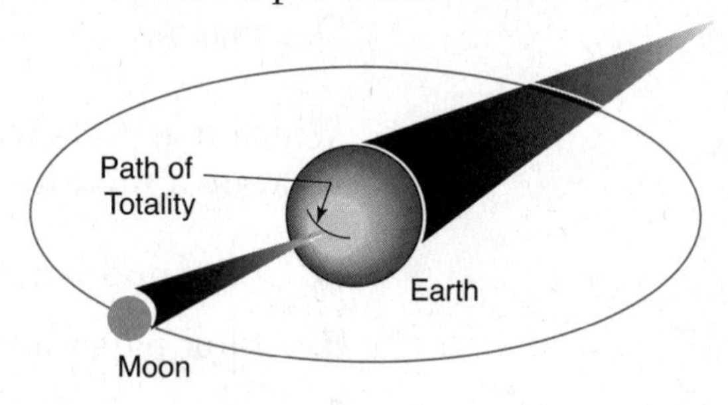

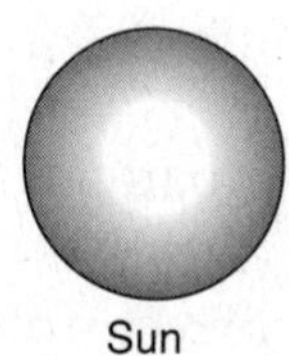

    A. when the Moon is between the Sun and Earth.
    B. when the Earth is between the Moon and Sun.
    C. when the Sun is between the Earth and Moon.
    D. during the Full Moon phase.

14. Which form of electromagnetic radiation causes skin damage?

    A. visible light
    B. ultraviolet
    C. infrared
    D. radio waves

15. Which layer in the soil diagram contains the most organic matter?

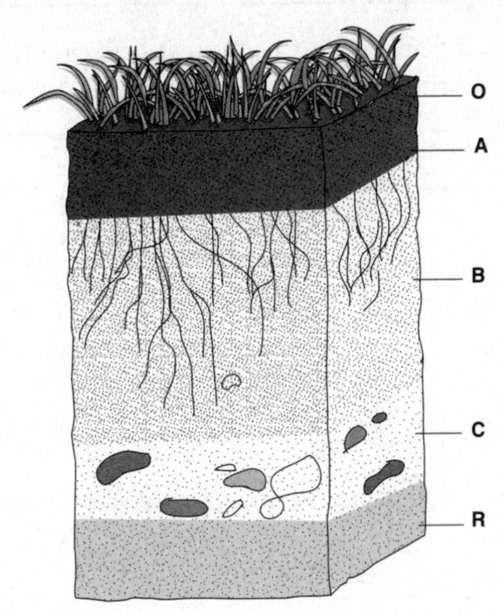

A. A

B. B

C. C

D. O

Respond fully to the open-ended question that follows. Show your work and clearly explain your answer. You may use words, tables, diagrams, or drawings.

---

16. Lizards are adapted to live in the desert habitat. Name and describe three adaptations that help lizards survive in the desert.

PART 2

For each of the questions or incomplete statements below, choose the best of the answer choices given.

17. Which of the following is a renewable resource?

A. coal
B. soil
C. solar energy
D. natural gas

18. The Moon revolves around

A. the Sun.
B. the Earth.
C. the solar system.
D. a comet.

19. Which of the following materials is opaque?

A. frosted glass
B. wax paper
C. a wood door
D. clear glass

20. Which process makes food for plants?

A. reproduction
B. precipitation
C. photosynthesis
D. respiration

21. Which of the following processes occurs in the formation of igneous rocks?

A. lithification

B. weathering

C. crystallization

D. metamorphism

22. Which is an example of a chemical reaction?

A. the freezing of water

B. the burning of wood

C. the tearing of a piece of paper

D. the shattering of glass

23. Which part of the cell contains most of the cell's genetic material?

A. cell membrane

B. nucleus

C. cytoplasm

D. mitochondria

24. Meteorologists measure air pressure using

A. thermometers.

B. anemometers.

C. barometers.

D. the Richter scale.

25. Which types of characteristics can be inherited?

A. characteristics that result from exposure to the environment

B. characteristics that are a result of diet and exercise

C. characteristics controlled by genes

D. characteristics produced by accident

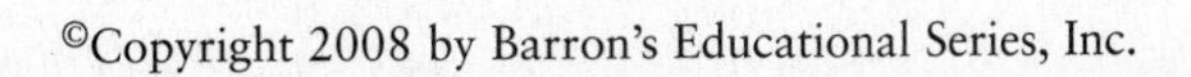

26. More solar energy reaches and warms regions near the equator because

A. it receives more direct sunlight.

B. it has more vegetation.

C. it is closer to the Sun.

D. it has more bodies of water adjacent to land.

27. Which of these demonstrates predation?

A. a flea living off a dog

B. a cow eating grass

C. a fox hunting a rabbit

D. a tick biting a human

28. Which of the following could be used to demonstrate the transformation of energy from electricity to heat?

A. barometer

B. a toaster

C. a pair of eyeglasses

D. a gas grill

29. Which plant cell part has the function of absorbing light energy and making food?

A. chloroplasts

B. vacuoles

C. cytoplasm

D. cell wall

30. How long does it take the Earth to complete one revolution around the Sun?

A. one year

B. one season

C. one month

D. one day

31. An owl hunting a rodent is an example of which kind of relationship?

A. parasite/host
B. consumer/producer
C. predator/prey
D. consumer/decomposer

Respond fully to the open-ended question that follows. Show your work and clearly explain your answer. You may use words, tables, diagrams, or drawings.

---

32. The Earth's climate may be getting warmer because of the actions of humans. List two human activities that may contribute to warming of the Earth's climate and suggest one way that the human race can reduce its impact on the Earth's climate.

PART 3

For each of the questions or incomplete statements below, choose the best of the answer choices given.

---

33. A mixture of iron filings and sand can be separated by

A. pouring the mixture through filter paper.
B. performing a chemical reaction.
C. adding heat to the mixture.
D. using a magnet.

34. Which of the following provides evidence of the theory of plate tectonics?

A. ocean currents
B. weather patterns
C. seafloor spreading
D. seasons

35. Animals that consume other animals are called

A. carnivores.

B. decomposers.

C. herbivores.

D. producers.

36. A burning candle demonstrates energy being transformed

A. from chemical energy to nuclear energy.

B. from heat energy to mechanical energy.

C. from light energy to nuclear energy.

D. from chemical energy to heat and light energy.

37. Which of these is not a natural component of soil?

A. humus

B. sand

C. plastic

D. clay

38. Which of these organ systems functions to continue the species?

A. respiratory

B. reproductive

C. nervous

D. digestive

39. Which of the following materials is transparent?

A. dictionary

B. wax paper

C. frosted glass

D. clear glass

40. Susie uses a compound microscope to observe cells. The eyepiece has a magnification of 10X and the objective lens has a magnification of 30X. What is the total magnification used for viewing the cells?

A. 100X

B. 300X

C. 10X

D. 30X

41. Which of these pictures illustrates how light is reflected off a mirror?

A

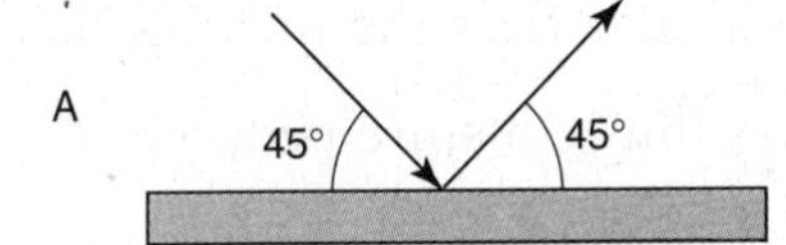

B

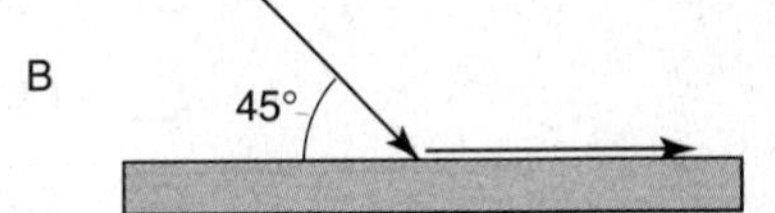

C

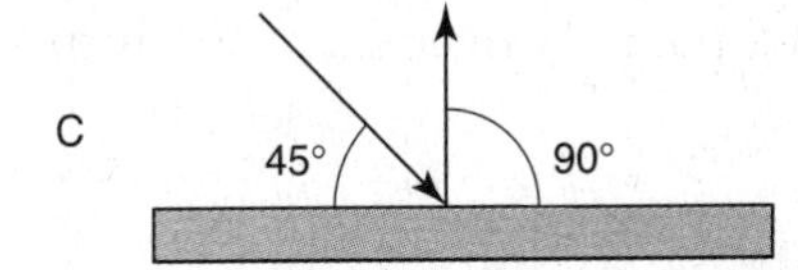

D

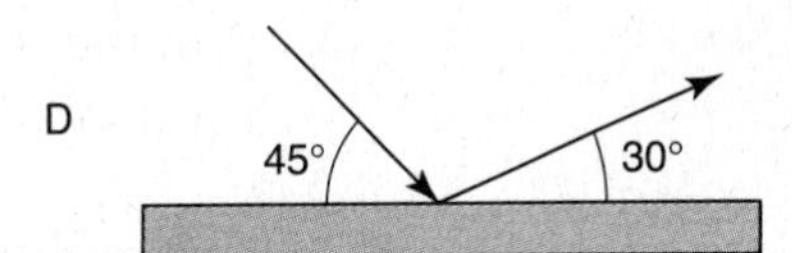

42. Which statement best describes the motion of the Earth?

A. It rotates as it revolves around the Sun.

B. It rotates as it revolves around the Moon.

C. The Earth rotates at the same speed that it revolves around the Sun.

D. The Earth revolves around both the Moon and the Sun.

43. What is the most basic unit of living things?

A. cells

B. tissues

C. organs

D. organ systems

44. Atoms are composed of protons, neutrons, and electrons. What is the electric charge of the proton?

A. positive

B. negative

C. neutral

D. double-charged

45. Which of the following questions cannot be answered through a scientific investigation?

A. How does air temperature affect air pressure?

B. How does salt affect the melting point of ice?

C. How does music inspire people?

D. What is the relationship between elevation and air pressure?

46. Johnny has blue eyes but both of his parents have brown eyes. If brown eyes is the dominant trait, describe the genetic makeup of his parents necessary for him to inherit the genes for blue eyes.

A. Both parents are purebreds.

B. Both parents are hybrids.

C. The female parent is a purebred, and the male parent is a hybrid.

D. The male parent is a purebred, and the female parent is a hybrid.

47. What is a similarity between sound waves and seismic waves?

A. They both carry energy.

B. They both travel through empty space.

C. They travel at the same speed.

D. They give off heat.

Respond fully to the open-ended question that follows. Show your work and clearly explain your answer. You may use words, tables, diagrams, or drawings.

48. The Sun is much larger than the Moon, yet from the Earth they appear almost the same size. Explain why this is. Explain and diagram the positions of the Earth, Moon, and Sun that can cause solar and lunar eclipses.

PART 4

For each of the questions or incomplete statements below, choose the best of the answer choices given.

49. The organisms that interact in a local park include birds, insects, trees, spiders, grass, and squirrels. Taken together, the organisms compose

A. a population.

B. a community.

C. a species.

D. a kingdom.

50. The gravitational force between the Earth and an object depends on the object's mass and

A. the object's distance from Earth.

B. the object's volume.

C. the object's state of matter.

D. the object's shape.

51. Air is made of a mixture of gases. Which gas is found in the greatest abundance in our atmosphere?

A. Nitrogen

B. Oxygen

C. Carbon dioxide

D. Hydrogen

52. Which two human body organs work together to make sure oxygen is transported to other areas of the body?

A. heart and lungs
B. stomach and heart
C. heart and stomach
D. lungs and stomach

53. Objects resist any change in their motion due to their

A. inertia.
B. position.
C. gravity.
D. chemical composition.

54. Food, gasoline, and batteries store what kind of energy?

A. mechanical energy
B. electrical energy
C. nuclear energy
D. chemical energy

55. The data table below shows the outdoor temperature at different times during one day. When was the highest temperature recorded?

Time	Temperature
6:00 A.M.	45°F
8:00 A.M.	49°F
10:00 A.M.	52°F
12:00 P.M.	58°F
2:00 P.M.	62°F
4:00 P.M.	57°F
6:00 P.M.	52°F

A. 6:00 A.M.
B. 12:00 P.M.
C. 2:00 P.M.
D. 6:00 P.M.

56. Which of the following is an example of a population?

A. mice and rats living in an alley
B. ants in a colony
C. trees, shrubs, and grasses in a field
D. cat living in a house with a family

57. Which of the following is not a mixture?

A. air
B. milk
C. pancake mix
D. salt

58. What process creates food, in the form of sugar, for plants?

A. evaporation
B. photosynthesis
C. respiration
D. oxidation

59. Based on the table below, which temperature is the best forecast for 6:00 P.M.?

Time	Temperature
12:00 P.M.	63 degrees Fahrenheit
2:00 P.M.	57 degrees Fahrenheit
4:00 P.M.	54 degrees Fahrenheit
6:00 P.M.	?

A. 56 degrees Fahrenheit
B. 60 degrees Fahrenheit
C. 54 degrees Fahrenheit
D. 52 degrees Fahrenheit

60. When plants and animals die, what happens to the nutrients and chemical energy stored within the plant and animal matter?

A. They are returned to the environment by decomposers.
B. They disappear when the organism dies.
C. They are absorbed into the atmosphere.
D. They are transferred to another ecosystem.

61. Some people take vitamin supplements, like calcium, to build strong bones. The mass of the calcium in these vitamin supplements is measured in

A. milligrams.

B. millimeters.

C. milliliters.

D. milliseconds.

62. Animals that eat plants are called

A. carnivores.

B. decomposers.

C. herbivores.

D. producers.

63. If two equal forces act on an object from opposite directions, the forces are said to be

A. gravitational.

B. universal.

C. balanced.

D. unbalanced.

Respond fully to the open-ended question that follows. Show your work and clearly explain your answer. You may use words, tables, diagrams, or drawings.

---

64. Constructive and destructive forces create and change landforms over time. Provide one example of a constructive force and one example of a destructive force. Explain how these forces could affect a mountainous area over time.

ANSWER KEY

PRACTICE QUIZZES

CHAPTER 1

1. **C** This response is an opinion question and cannot be tested through an investigation.
2. **B** When the line stays straight and horizontal there is no change in the temperature as time passes.
3. **C** The temperature of each dish is changed purposely and is therefore the independent (manipulated) variable.
4. **C** This graph shows a direct relationship between number of chirps per minute and the temperature.
5. **B** Response A is a conclusion statement, response C is written in the hypothesis format (if . . ., then . . .) but is illogical, and response D is a problem written as a question. Response B is a hypothesis written in the "if . . ., then . . ." format and shows the predicted relationship between eating breakfast and grades in school.
6. **C** A graduated cylinder and water can be used to perform the water displacement method of determining the volume of an irregularly shaped solid.
7. **D** Response D is an inference based on observations, not an observation.
8. **B** In 2003, the rodent population is the highest, and barn owls would not have to compete as much because of the abundance of rodents.
9. **C** Both liquids evaporated, but rubbing alcohol evaporated faster because there was less left a day later.

10. C Using interpolation of the data, the plant grows about 0.2 centimeters each day. Hence, on Sunday the height would be 1.2 centimeters.

11. A When similar data is collected when an experiment is repeated, it confirms that the data is valid.

12. C Response C is a prediction written in the "if . . ., then . . ." format. Responses A and D are questions. Response B is a knowledge statement.

13. B To find the height of a 1:1,000 scale model, divide the actual height of the volcano by 1,000.

14. C At 2:00 P.M. the temperature was 62 degrees Fahrenheit, which is the highest temperature recorded during data collection.

15. A A triple beam balance measures mass. A spring scale measures the force of gravity on an object. A thermometer measures temperature. A graduated cylinder measures the volume of a liquid.

16. D Air conditioning and refrigeration allow for foods to be preserved for longer periods of time. They can then be distributed around the country and world in refrigerated trucks and airplanes.

17. C Using extrapolation, you can predict that the snail will travel a distance of 6 centimeters in 6 minutes.

18. A Binoculars can be used to view objects and organisms that are far away and difficult to see with the naked eye.

19. B The elephant has a life span of 70 years, which is closest to that of a human's, that is, 77 years.

20. B A graduated cylinder should be read at the level of the liquid it is containing.

CHAPTER 2

1. C Gills and lungs are both respiratory organs, which are used to obtain oxygen.

2. D Organisms classified in the same species have few differences in structure.

3. C Only characteristics controlled by genes can be passed to the next generation. The other characteristics are acquired and cannot be passed to offspring.

4. B The reproductive system functions to produce offspring; this continues the existence of the species.

5. C Bacteria are prokaryotes because they do not have nuclei (plural of nucleus) in their cells.

6. B Chromosomes are found in the nucleus of a cell, and they carry the genetic information.

7. D Cells are the basic structural units, or building blocks, of living things.

8. A Only plant cells have both cell walls and chloroplasts.

9. B An increase in predators would probably cause the dodo population to decrease. An increase in food supply and an increase in nesting locations would help the dodo survive. The dodo should not be affected by the climate if it is stable and not changing.

10. B Photosynthesis occurs in plants and creates food, in the form of sugar, from the raw materials of water, carbon dioxide, and sunlight.

11. A Nerve cells transmit electrical signals from all areas of the body to the brain. This requires a lot of energy. Mitochondria provide the cell with energy.

12. A All of these organisms are in the animal kingdom.

13. C In sexual reproduction, half of the genes come from each parent; the offspring have a mixture of traits from their parents.

14. B Red blood cells carry oxygen to all parts of the human body.

15. D A child inherits traits from both his mother and his father.

16. D Arthropods are a phylum that is classified within the animal kingdom.

17. B The cell membrane allows nutrients and other materials to enter and exit the cell.

18. B The only chance that two black-furred guinea pigs could have an offspring with brown fur is if both parents are hybrids having the genotype Bb; both parents must pass down the recessive trait (b) for an offspring to be brown.

19. A Fungi obtain food through absorption. All of the other answers are false.

20. Rubric for Open-Ended Question—A *complete response* includes the following: If both parents are hybrids and carry both the gene for brown hair and the gene for blond hair, then there will be a 25% chance that their child will have blond hair. Both parents will have to pass the gene for blond hair in order for their child to have blond hair. (B—brown hair, b—blond hair)

	B	b
B	BB	Bb
b	bB	bb

A *partial response* is lacking in detail or does not include a Punnett square.

An *incomplete response* incorrectly explains how a blond-haired child could be produced.

CHAPTER 3

1. A A maple tree is a plant and is therefore a producer, which uses energy from the Sun to make food.

2. B Humans are consumers because we depend on plants and animals for food and energy.

3. B Tropical rainforests, with their high temperatures and high precipitation, contain the most diversity of life.

4. A Algae is a producer, which uses energy from the Sun to make food.

5. A Decomposers return the nutrients and energy found in dead plant and animal matter to the environment.

6. D The lion is the predator of the zebra, and the zebra is the prey.

7. D Decomposers absorb nutrients from decaying plants and animals.

8. C Since the barn owl preys on mice, when there is an increase in the mouse population there will be less competition between owls for food. This will cause the barn owl population to increase since there is more food to support more owls.

9. C The energy in producers is transferred to herbivores and then to carnivores. Herbivores consume producers. Carnivores consume herbivores.

10. Rubric for the Open-Ended Question—A *complete response* includes that

1. ozone depletion is due to the release of chemicals called CFCs into the atmosphere
2. the destruction of ozone will cause more ultraviolet radiation from the Sun to reach the earth's surface
3. exposure to this UV radiation can cause skin cancer in humans.

A *partial response* includes only one of the responses above.

An *incomplete response* shows no understanding of the ozone layer and its relationship to UV radiation and skin cancer.

11. D All of these animals are found in a tropical rainforest.

12. A Oil is a fossil fuel and a nonrenewable resource.

13. A Koalas have only one food source, eucalyptus. If that food source became depleted, then koalas would not survive. All of the other marsupials in the table have more than one food source. If one of their food sources is depleted, then those marsupials would have other options to consume.

14. B A tick is a parasite.

15. C Bacteria are decomposers.

16. A Spiders prey on insects; if the spider population increases, then there will be more spiders to consume insects. Eventually, this could cause a decrease in the insect population.

17. A Through the process of photosynthesis, solar energy is converted to chemical energy. This chemical energy is stored in plants. As other organisms consume the plants, the energy is transferred through the food web.

18. C Permafrost, or permanently frozen soil, is found in the tundra.

19. A Algae is a green plant and is therefore a producer. Minnows eat algae and are called consumers.

20. Rubric for the Open-Ended Question—A *complete response* includes any two of the following possible answers:

- The hawks would have to rely on rodents as their only food source; therefore, the hawk population might decrease because of increased competition for food.
- The frog population will increase because the frog will not have any predators.
- The caterpillar population will decrease because there will be more predators (the increase in the frog populations).

A *partial response* includes only one of the possible answers above.

An *incomplete response* incorrectly predicts how other organisms would be affected.

CHAPTER 4

1. B Both objects A and B have a density of 2 g/cm^3, while object C has a smaller density of 1 g/cm^3.

2. C Neutrons have a neutral charge. Protons have a positive charge and electrons have a negative charge.

3. A There is chemical energy stored in batteries. The batteries are connected to wires, which convert the chemical energy to electricity. The light bulb converts the electricity to light energy.

4. C Gamma rays have the shortest wavelength, while radio waves have the longest wavelength.

5. A A nail rusting is a chemical reaction. The other three changes are all physical changes because the composition of the substance stays the same, but the shape, phase, or location may have been altered.

6. B Amplitude is the height of the wave. Trough is the lowest point of the wave. Wavelength is the length of the wave between two adjacent corresponding points. Frequency is the number of waves in a period of time.

7. D Clear glass is transparent because it transmits light. Wax paper and frosted glass are translucent because they transmit some light but scatter other light. A dictionary is opaque because it absorbs or scatters all of the light that strikes it.

8. D A skier going downhill uses the force of gravity for movement.

9. C Electrical energy is easy to transport and is commonly used in households because it travels through wires internally made of conductive materials.

10. B Convection currents are the upward movement of warm, less dense air and the sinking of cooler, denser air. Hot air balloons use convection to rise up in the air.

11. D H_2O contains three atoms, two atoms of hydrogen and one atom of oxygen. Choice A—O_2 contains two atoms of oxygen. Choice B—NaCl contains one atom of sodium and one atom of chlorine. Choice C—N_2 contains two atoms of nitrogen.

12. C Gold is an element and elements cannot be made or changed by ordinary means.

13. B Density equals mass divided by volume.

14. B When a liquid changes to a solid, the molecules seem to lose energy. They slow down and move closer together.

15. D Diagram D shows the wave with the longest wavelength. The wavelength is measured between adjacent corresponding position of the wave.

16. A A roller coaster speeding down a hill converts the potential energy from its height to kinetic energy as its velocity increases as it moves down the hill. All of the other responses are examples of objects with stored potential energy.

17. A According to the law of universal gravitation, gravity is a force that exists between objects and depends on the mass of the objects and the distance between them.

18. **D** Sand and iron filings can be separated using a magnet because iron has the physical property of magnetism, but sand does not.

19. **C** Oxygen is an element. In its gas form, it is composed of two atoms of oxygen bonded together. Air is a mixture of many gases. Water is a compound consisting of hydrogen and oxygen. Paper is a mixture composed of a variety of compounds and other substances.

20. Rubric for Open-Ended Question—A *complete response* includes the following possible procedures and supplies as indicated:

    - Because wood is less dense than water, it will float on the surface. Skim the wood particles off the surface of the water using a screen.
    - Pour the sugar water with gravel and iron filings through filter paper in a funnel. The sugar water will come through the filter because the sugar is dissolved in the water, but the gravel and iron filings will remain on the filter paper.
    - Boil the sugar water or allow the water to evaporate so that you are left with the sugar crystals.
    - Use a magnet to separate the iron filings from the gravel. The iron filings are magnetic, but the gravel is not.

    A *partial response* is missing one of the steps or does not explain the supplies needed or properties used.

    An *incomplete response* incorrectly explains the separation process or just lists the supplies.

CHAPTER 5

1. **B** As elevation increases, the density of the air decreases. There is less oxygen gas when air has a lower density.

2. **C** Marble is a metamorphic rock.

3. **B** In the Earth-Moon-Sun system, the Earth revolves around the Sun, and the Moon revolves around the Earth. The Sun is at the center of this system.

4. **C** A seismograph records seismic waves; a telescope is used to magnify objects that are far away. A barometer measures air pressure. A thermometer measures temperature.
5. **A** As water freezes in cracks in rocks, the rock will break apart, which is physical weathering.
6. **B** Lithospheric plates move over the asthenosphere. Convection currents in the asthenosphere cause the plates to slowly move.
7. **B** Uranus is a gas giant. Mercury, Venus, and Earth are terrestrial planets.
8. **C** As the Moon revolves around the Earth, different areas of the moon that are illuminated by the Sun are observed.
9. **D** Lunar eclipses occur when the Earth is between the Sun and Moon; the Earth's shadow blocks the Sun from illuminating its surface. Lunar eclipses occur during a Full Moon.
10. **B** June 21st is the summer solstice, or the longest day of the year. At this time, the northern hemisphere is tilted toward the Sun and receives more direct sunlight.
11. **A** Telescopes are used to magnify objects that are far away, making them easier to observe.
12. **D** Moh's scale is used to determine the relative hardness of minerals. The Richter scale is used to determine the magnitude of an earthquake. The Beaufort scale is used to measure wind. The Fahrenheit scale is used as a temperature scale.
13. **B** Fossils are found in sedimentary rocks. The heat that is involved with forming igneous and metamorphic rocks destroys any evidence of fossils.
14. **C** The core of the Earth is the hottest layer.
15. **A** The third quarter phase is the next phase in the sequence of moon phases.
16. **B** The Sun is a star.
17. **A** Physical weathering is caused primarily by wind and water, breaking down the rocks into smaller pieces. Chemical weathering is caused by chemical reactions and changes the chemical composition of the rocks.

18. A The moon is illuminated by the Sun, and we can see the Moon because it reflects light from the Sun.

19. C The Moon revolves around the Earth in a period of time roughly equal to one month, or about 29 days.

20. Rubric for Open-Ended Question—A *complete response* includes a clear and thorough explanation of evaporation, condensation, precipitation, and runoff.

A *partial response* is missing one of the steps, or explains the steps but does not name them scientifically.

An *incomplete response* incorrectly explains the water cycle or merely lists some of the steps with little or no explanation.

ANSWERS: PRACTICE TEST 1

Part 1

1. A A mosquito is a parasite and when it bites a person, the person becomes the host.

2. A Carbon dioxide is made of two different elements; the other choices are made of only one element.

3. B As the Earth rotates, different hemispheres face the Sun, receiving daylight.

4. B The asthenosphere is located just below the lithospheric plates. The convection movements in this layer cause the plates to move.

5. D The scattering of light is the bouncing off of light rays. Refraction is the bending of light as it transmits. Transmission occurs when light passes through an object.

6. B A land breeze occurs in the evening; sea breeze occurs during the day.

7. D The color of the insect helps to camouflage it in its habitat.

8. B Magnesium has an atomic number of 12 and has twelve protons.

9. C It takes the Moon about one month or 29 days to complete its revolution around the Earth.

10. C A water lily is the only plant among the choices. Green plants undergo photosynthesis.

11. B Electrons have a negative charge.

12. A The moon reflects light from the Sun.

13. B Cells are the simplest; organ systems are the most complex.

14. C The color of an opaque object is the color of light the matter scatters.

15. D Bacteria are decomposers.

16. Rubric for Open-Ended Question—A *complete response* could include the following possible materials and procedures: When coal, oil, or natural gas is burned, the chemical energy in the fossil fuel is transformed into heat energy. This heat energy is used to boil water. The steam rising from the boiling water turns a fan-like piece of equipment called a turbine; the energy has been converted to mechanical energy. The spinning turbine is connected to an electric generator. A generator consists of a long coiled wire surrounded by a magnet. As the generator turns, an electric current is produced in the wires, converting the mechanical energy to electricity. The electricity then travels through power wires to homes and buildings to provide us with energy for running our appliances. Alternatively, sources such as wind or tidal currents could also be used to spin turbines and generators to produce electricity.

A *partial response* is missing one of the steps or does not explain the equipment needed or transformations.

An *incomplete response* incorrectly explains the process of energy transformations to harness energy.

Part 2

17. B Sedimentary rocks can contain fossils.

18. A Frosted glass transmits some light and scatters some light. Glass is transparent, and answers B and C are opaque.

19. A Digestion primarily takes place in the stomach and small intestine.

20. D Canyons are created by erosion by wind and water. The three other choices are all created by constructive forces.

21. D Soup mix is a mixture of various compounds.

22. A Algae is a producer and converts sunlight into food through photosynthesis.

23. D December 21st is the first day of winter. On this date, the northern hemisphere is tilted away from the Sun, and we experience shorter days.

24. D This is an inference based on observations, but not an observation.

25. C Nerves send messages from all over the body to the brain.

26. A Lithification is the cementation of sediments into rock.

27. A The sex cells, sperm and egg, contain genetic information that combines to form a new organism.

28. C The rotation of the Earth causes day and night.

29. B When data are replicated, their validity is confirmed.

30. C The movement of the earth's crust at faults causes earthquakes.

31. C Diagram 3 shows the refraction or bending of light as it reaches and transmits through water.

32. Rubric for Open-Ended Question—A *complete response* could include the following possible materials and procedures: Various setup methods and pieces of equipment would work. The answer must include the idea that water must be evaporated, leaving behind the salts originally found in the seawater. The water vapor needs to be condensed and collected in a container so that it can be used for drinking water. The response could also include an explanation of distillation.

A *partial response* is missing one of the steps or does not explain the equipment needed or forgets steps like the condensation of the liquid.

An *incomplete response* incorrectly explains the process of acquiring fresh water from salt water.

Part 3

33. A Populations consist of just one species in an area, community refers to all of the living organisms, and ecosystems are all of the organisms and their nonliving surroundings.

34. C Because the Earth is tilted, different hemispheres of the Earth receive varying amounts of direct sunlight through the year.

35. B This is the only relevant hypothesis, or prediction.

36. D Both C and D are good answers; however, D is more specific because it draws a conclusion from the observations.

37. A Red blood cells carry oxygen.

38. A Solar energy drives the water cycle.

39. C The temperature of the dishes was the variable purposely changed.

40. B Protons and neutrons are located in the nucleus of atoms.

41. B Neptune is a gas giant. The other selections are terrestrial planets.

42. A The first organism in a food chain is always a producer.

43. C C is a prediction, or hypothesis. A and D are questions, and B is a fact statement.

44. B Meiosis forms sex cells.

45. A Nitrogen makes up 78% of the atmosphere.

46. C If the grasshoppers were killed by insecticides, frogs would not have as much food and some would die, decreasing the population.

47. C Wavelength is the length between two corresponding points of adjacent waves.

48. Rubric for Open-Ended Question—A *complete response* could include the following possible materials and procedures: The student would need a stopwatch. First she should measure her resting heart rate. Then she should exercise for a given amount of time. After she stops, she should record her heart rate at regular intervals until it comes down to the resting heart rate.

A *partial response* is missing one of the steps or does not explain the equipment needed.

An *incomplete response* incorrectly explains the process of recording and tracking heart rate.

Part 4

49. A A niche is an organism's role in an ecosystem.

50. A Children have traits from both parents since their genetic makeup is from both parents.

51. C As the Moon revolves around the Earth, we can view changing portions of the Moon's illuminated surface.

52. B The arrows are pointing to electrons in the diagram.

53. C Cows have an average lifespan just two years longer than dogs.

54. B Evaporation caused by the solar energy changes the liquid water to the gas form.

55. C The statement is simply an observation.

56. A If she only waters half of the plants then she will have a control group with which to compare her results.

57. D Fossil fuels are from the remains of organisms and are associated with sedimentary rocks.

58. C Sex cells contain half the number of chromosomes as body cells.

59. C The particles in the picture appear attached to other particles. A and D are elements and B is a mixture.

60. C All insects have six legs.

61. D A water molecule contains three atoms, two hydrogen and one oxygen.

62. A Tissues are composed of similar cells.

63. C In the process of evaporation, liquid water molecules gain energy and turn to a gas.

64. Rubric for Open-Ended Question—A *complete response* could include the observation that microscopes allow doctors to determine whether people are healthy by examining specimens of the patient's cells; then list at least one other tool that helps doctors and one way a doctor's job would be different if he or she did not have these tools.

A *partial response* is lacking in detail and only answers some of the questions.

An *incomplete response* fails to answer the question with detail or accuracy.

ANSWERS: PRACTICE TEST 2

Part 1

1. **A** The rotation of the Earth causes the deflection of global winds.
2. **B** Humans are consumers because we consume plants and animals.
3. **C** Potential energy converts to kinetic energy as an object loses height and its velocity increases.
4. **B** Tropical rainforests contain a wide variety of organisms.
5. **B** Diagram B is the most accurate model of the diagrams presented.
6. **A** A string vibrating on a guitar produces a sound wave.
7. **B** The cell membrane allows for materials to pass into and out of the cell.
8. **A** Igneous rocks form from the cooling of molten rock.
9. **B** Water could be mixed with the sand and sugar. The sugar will dissolve in the water. Pour the mixture through the filter paper. The sand will remain in the filter paper and the sugar can be evaporated out of the sugar water.
10. **C** The arrow represents the flow of energy from producers to herbivores and then to carnivores.
11. **A** Day and night are caused by the rotation of the Earth.

12. D First lichen and moss will grow, then grasses, scrub growth, and eventually pine trees.

13. A The Moon is between the Sun and Earth blocking out the light from the Sun for a few minutes.

14. B Ultraviolet light is harmful to the skin.

15. D Horizon O is made of mostly humus, or organic matter.

16. Rubric for Open-Ended Question—A *complete response* could include and thoroughly describe any three of the following: cold-blooded, thick, scaly skin to protect it from the hot dry climate and hold in moisture; claws for moving through the sand; long tail for balance and that can be regenerated if attacked by predator; color that is easy to camouflage in habitat; and nocturnal.

A *partial response* does not list and explain three of the lizard's features.

An *incomplete response* incorrectly explains adaptations of the lizard.

Part 2

17. C Solar energy is renewable; we can't use it up.

18. B The Moon revolves around the Earth.

19. C A wood door absorbs and scatters the light that strikes it.

20. C Photosynthesis is the process whereby plants make food.

21. C The cooling, solidification, and crystallization occurs in the igneous rock formation process.

22. B The burning of wood creates new chemical products.

23. B The nucleus contains chromosomes made of genetic material.

24. C Barometers are used to measure air pressure.

25. C Only traits you were born with can be passed on to the next generation.

26. A The Sun is directly overhead near the equator.

27. C The fox is the predator, and the rabbit is the prey.

28. B A toaster converts electricity to heat.

29. A Chloroplasts absorb light energy and make food.

30. A The Earth travels around the Sun in one year.

31. C The owl is the predator, and the rodent is the prey.

32. Rubric for Open-Ended Question—A *complete response* could include the following: Humans have polluted the atmosphere with various chemical gases and particles. Humans have also cut down forests, which are part of the dynamic ecosystem, and polluted waterways. Humans can try to reduce their amount of waste and recycle materials. They can also carpool or use mass transportation so as not to contribute to air pollution.

A *partial response* is missing one of the components of the response.

An *incomplete response* incorrectly explains the human impact on the warming climate.

Part 3

33. D A magnet can be used to remove the iron filings.

34. C Sea floor spreading provides evidence of plate tectonics.

35. A Carnivores are animals that eat other animals.

36. D The chemical energy stored in the candle is released as heat and light energy in the flame.

37. C Sand, silt, clay, and humus compose soil, NOT plastic.

38. B The reproductive system functions to produce offspring and thus continue the species.

39. D Clear glass is transparent; you can see right through it.

40. B 30X × 10X = 300X.

41. A Light bounces off at the same angle it approached the mirror.

42. A The Earth is rotating all the while it is revolving around the Sun.

43. A Cells make up living things.

44. A Protons have a positive charge.

45. C This question is a matter of personal opinion, not science.

46. B If both of Johnny's parents are hybrid for eye color, then they would have a 25% chance of having a blue-eyed child.

47. A Sound and seismic waves carry energy.

48. Rubric for Open-Ended Question—A *complete response* includes the fact that the Sun is extremely far away from the Earth and that the Moon and the Sun appear to be the same size because of this distance. Solar and lunar eclipses can occur when the Sun, Earth, and Moon are aligned. When the Moon is between the Sun and Earth, there is a solar eclipse, and when the Earth is between the Sun and Moon, there is a lunar eclipse.

A *partial response* does not fulfill all components of the prompt.

An *incomplete response* incorrectly explains the processes.

Part 4

49. B A community is all the organisms in an area.

50. A Gravity is due to mass and distance between bodies.

51. A Nitrogen makes up 78% of the Earth's atmosphere.

52. A The heart and lungs pump oxygen around the body.

53. A Inertia is an object's tendency to resist motion.

54. D These substances store chemical energy.

55. C At 2 P.M. the temperature was at a high of 62 degrees Fahrenheit.

56. B A population consists of one species.

57. D Salt is not a mixture; it is a compound.

58. B Photosynthesis creates food in the form of sugar for plants.

59. D The temperature appears to be lowering.

60. A Matter from dead organisms is returned to the environment by decomposers.

61. A Vitamin supplements are measured in milligrams.

62. C Herbivores are animals that consume plants.

63. C When forces are equal and in opposite directions, they are balanced.

64. Rubric for Open-Ended Question—A *complete response* could include any of the following: Constructive forces are volcanic eruptions and also plate boundaries where plates converge. Destructive forces are weathering and erosion. These forces can affect a mountainous area by continuing to build it up over time through tectonics; and weathering and erosion could wear down the surface of the mountain over time.

A *partial response* is lacking in detail and only answers some of the questions.

An *incomplete response* fails to answer the question with detail or accuracy.

GLOSSARY

abiotic – the nonliving parts of the environment.
absorption – to retain or take in light energy as it strikes a substance.
acceleration – any change in the velocity, or the speed and direction, of an object.
accuracy – how close a measurement is to its true value.
acid rain – acidic precipitation due to industrial pollution, which can damage plants, disturb water ecosystems, and damage structures.
acquired traits – characteristics of an organism that are a result of interactions with the environment.
air mass – a large region of air with uniform characteristics, such as temperature and pressure.
allele – a form of a gene.
amplitude – the height of the wave from the starting undisturbed position.
animals – one of the six kingdoms of living things; multicellular eukaryotes that are heterotrophs by ingestion, such as jellyfish, worms, insects, and mammals.
archaebacteria – one of the six kingdoms of living things; ancient unicellular bacteria that are also prokaryotes and autotrophs.
asexual reproduction – a form of reproduction in which a single cell divides into two daughter cells through mitosis. The offspring shares identical characteristics with its parent.
asteroid belt – the region in our solar system found beyond the four terrestrial planets that is composed of many asteroids and the dwarf planet Ceres. The asteroid belt is located between the orbits of Mars and Jupiter.
asthenosphere – the partially melted, deformable rock in the upper mantle of the Earth upon which the lithosphere floats.
atmosphere – a mixture of gases surrounding the Earth and consisting mostly of nitrogen and oxygen gases.
atom – the fundamental particles of an element that have all the chemical and physical properties of the element.

atomic number – the number assigned to an element that equals the number of protons in an atom of that element.
atomic weight – the average mass of an atom of an element.
autotroph – an organism that produces its own food from the sunlight and other nutrients.

balanced forces – forces that will not cause a change in the speed or direction of an object's motion.
bar graph – a visual representation of descriptive, qualitative data.
barometer – an instrument used to measure air pressure.
basalt – a type of igneous rock that usually forms at mid-ocean ridges and composes the ocean floor.
best-fit line – a line drawn on a line graph to represent the general pattern of data when the actual data points cannot be connected with a straight line.
bilateral symmetry – symmetry found in organisms like mammals. Organisms have right and left sides that are mirror images of each other.
biodiversity – the number and diversity of organisms in an ecosystem.
biomass – the total mass of all the organisms in a particular population or in a given area.
biome – large ecosystems that extend over wide regions.
boiling point – the temperature at which a particular liquid boils at a fixed pressure.
brackish – a mixture of seawater and fresh water found in estuaries.

carnivore – consumer that feeds mostly on other animals.
carrion – dead and decaying animal matter.
cell – the basic unit of structure in organisms.
cell membrane – the thin, skin-like structure that surrounds the cell and acts as a barrier, allowing some materials, such as water, to enter and exit the cell.
cell wall – provides a framework, or rigid, tough structure, outside the cell membrane in plant cells.
central tendency – the average of a set of data, such as the mean, median, or mode.
chemical change – a change in matter in which two or more materials react and a new substance forms that has different properties than the original materials. Energy is often absorbed or released in a chemical change. Chemical change can also be called a chemical reaction
chemical energy – the form of energy that is stored in the bonds between atoms of compounds.

chemical equation – a representation or model of a chemical reaction that uses chemical symbols and formulas.
chemical formula – a symbol used to represent the number and type of atoms in a molecule of a particular compound.
chemical symbol – a one- or two-letter symbol for an element.
chemical weathering – the process of breaking down, or changing, the chemical composition of rocks; usually produces clay.
chlorofluorocarbons – chemical pollutants that were released into the atmosphere until the 1990s and that destroyed the ozone gas within the ozone layer in the atmosphere.
chloroplasts – the food producers of plant cells that contain a pigment called chlorophyll, which provide plants with their green color and play a role in the process of photosynthesis.
chromosome – made of the genetic material, DNA, of an organism, and is found within the nucleus of cells.
class – a grouping between phylum and order within the biological classification system (kingdom, phylum, class, order, family, genus, species), where kingdoms are broad, general groupings of organisms and species are more specific groupings.
climate – the general weather conditions in a given area over a long period of time, which is affected mostly by latitude, elevation, and location relative to large bodies of water and mountain ranges.
climax community – the final stage of succession.
coefficient – a number found before a chemical symbol or formula, indicating the number of atoms or molecules of each reactant and product that are involved in a chemical change.
cold front – the boundary that forms as a cold air mass pushes against and under a warm air mass.
comets – small, icy objects that travel around the Sun in highly elliptical orbits and that are made of dust, ice, and other chemical compounds.
commensalism – a symbiotic relationship in which one organism benefits while the other organism is neither affected nor harmed.
community – consists of only the living things in an ecosystem, or all of the populations in an ecosystem.
competition – the interaction between organisms where they strive to gain resources such as food, water, or territory.
compound – a substance made of two or more elements.
compound microscope – a microscope that uses more than one lens to magnify an object. There is a lens in the eyepiece and usually three lenses on a revolving nosepiece.
compression wave – a wave in which matter in the medium moves back and forth in the same direction of the wave; also called a longitudinal wave.

condensation – the process in which a gas changes into a liquid.

conduction – the direct transfer of heat from one material to another.

conductivity – the ability of a substance to transmit heat, electricity, or sound.

coniferous forest – the biome with long, cold winters, little precipitation that is mainly snow, and a landscape of evergreen trees.

constants – the variables in an experiment that stay the same; also called controlled variables.

consumer – an organism that cannot make its own food and must rely on other organisms as a source of food.

continental drift – an outdated theory that stated that the continents were once one large supercontinent, Gondwana, and have since drifted through the oceans to their current locations.

control – a parallel experiment that can be used for comparison and in which no variables have been manipulated, or changed.

controlled experiment – an experiment where only one variable is tested.

controlled fires – forest fires that are closely monitored and intentionally started, usually when winds are low, to renew forests and begin the succession process.

controlled variable – the variables in an experiment that stay the same; also called constants.

convection – the transfer of heat within a fluid.

convection currents – the movement of fluids in which as material warms, it becomes less dense and rises; then cooler, more dense material sinks in to take its place.

convergent plate boundaries – the location where two lithospheric plates collide.

Coriolis effect – the deflection, or curving, of winds due to the rotation of the Earth

crest – the topmost part of a wave.

crust – the very thin, outer layer of the Earth made of soil and solid rock.

cytoplasm – the jelly-like material that fills the cell, holds all of the cell parts in place, and helps to maintain the shape of the cell.

data – recorded observations.

data table – a chart with columns and rows, used to organize data.

decomposers – consumers that feed on dead and decaying plants and animals.

density – the amount of mass per unit volume of an object.

dependent variable – the variable that changes in response to the independent variable; also called the responding variable.
desert – the biome characterized by dry, often sandy regions, with high temperatures and little rainfall.
direct relationship – a relationship between variables in which they both increase or decrease; if one variable increases/decreases the other variable also increases/decreases.
divergent plate boundaries – the location where two lithospheric plates diverge, or move apart.
DNA – deoxyribonucleic acid; a long, twisted, thread-like molecule that holds the genetic code for building and maintaining an organism.
dominant – a stronger trait or gene.
ductile – the property of a substance to be drawn into long, thin wires; a physical property of metals.

Earth – an inner, terrestrial planet, which is the third planet from the Sun and supports life.
earthquakes – the movement of rock along a fault that often occurs near plate boundaries and releases seismic energy.
eclipse – phenomenon that occurs when the Earth, Moon, and Sun are aligned, blocking the light from the Sun.
ecology – the study of the interrelationships between animals and environments and how humans affect the environment.
ecosystem – consists of the living and nonliving things in a particular place that interact with each other.
electricity – the movement of electrons through a conductor.
electromagnetic energy – a form of energy that includes light, electricity, and electromagnetic radiation such as microwaves and gamma rays.
electromagnetic spectrum – the range of electromagnetic energy consisting of a variety of wavelengths and including gamma rays, x-rays, ultraviolet light, visible light, infrared light, and radio waves.
electrons – negatively charged particles that spin around the nucleus of atoms.
element – a simple substance that has characteristic physical and chemical properties.
energy – the ability to do work, or exert a force, over a distance.
environment – consists of all of the nonliving things around an organism, such as light, air, soil, landforms, temperature, water, and climate.
erosion – the movement of soil, rocks, and sediment by wind and water.

estuaries – a water biome located where the fresh water from rivers and streams meets the salt water from the ocean.

eubacteria – one of the six kingdoms of living things; unicellular bacteria that are prokaryotes, such as *E. coli* and *salmonella*.

eukaryote – an organism composed of a cell or cells that contain a nucleus.

evaporation – the process of a liquid changing to a gas.

evolution – the gradual change in a population of organisms that is the result of natural selection.

extinction – the permanent disappearance of a species of organisms.

extrapolation - estimating a value beyond known values.

family – a grouping between order and genus within the biological classification system (kingdom, phylum, class, order, family, genus, species), where kingdoms are broad, general groupings of organisms and species are more specific groupings.

fault – the location where there is movement of rock along a fracture or a break within rock.

fission – the splitting of atoms, which releases a tremendous amount of energy from the nucleus of atoms.

flooding – a rise and overflow of water onto land that is usually dry.

food chain – the transfer of matter and energy through organisms in an ecosystem.

food web – the interconnected food chains in an ecosystem that identify the feeding relationships between producers, consumers, and decomposers.

force – a push or a pull that can cause an object to move.

forest fire – a catastrophic natural event that consists of uncontrolled fires that destroy vegetation and can also damage homes and business structures.

friction – a force that acts in the direction opposite to the motion of the object; the force that resists motion when the surface of one object comes into contact with the surface of another object.

front – the boundary between two air masses.

fungi – one of the six kingdoms of living things; mostly unicellular eukaryotes that are also heterotrophs by absorption, such as molds, mildews, and mushrooms.

gas – the phase of matter that has no definite volume and no definite shape. The atoms or molecules of a gas will diffuse, or spread out, to fill whatever container it is in.

gene – sections of DNA that control a particular trait for an organism.

generator – a long, coiled wire surrounded by a magnet that, when spinning, produces an electric current in its wires, converting the mechanical energy to electricity.

genotype – actual genetic makeup for a particular trait.

genus – a grouping between family and species within the biological classification system (kingdom, phylum, class, order, family, genus, species), where kingdoms are broad, general groupings of organisms and species are more specific groupings.

geology – the study of the Earth.

global warming – the rising of overall surface temperature on Earth that can cause disastrous effects to many ecosystems.

global winds – winds that occur on a global scale due to differences in air temperature and pressure.

graduated cylinder – a cylindrical container with markings along the side, used to measure the volume of a liquid.

granite – a type of igneous rock that forms slowly below the Earth's surface.

graph – a visual representation of data, such as bar graphs and line graphs.

grasslands – the biome where the main vegetation is a variety of grasses; sometimes called savannas or prairies.

gravitational potential energy – stored energy due to position, or height.

gravity – a force that exists between objects that depends on the mass of the objects and the distance between them.

greenhouse effect – a process in which the atmosphere absorbs the Sun's radiation and keeps the planet warm.

groups – the columns on the Periodic Table that contain elements with similar properties.

habitat – the place where an animal or plant lives.

hardness – a physical property of minerals that can be observed and used to help identify a mineral.

heat energy – a form of energy due to the motion of atoms and molecules; also called thermal energy.

herbivore – consumer that feeds on plants.

heterotroph – an organism that cannot produce its own food and relies on other organisms for nutrients.

horizons – layers within soil that have different chemical compositions and textures.

host – the organism that is harmed within a parasitic relationship.

humus – dark, moist organic material made up of decomposed plants and organisms that contains nutrients needed by plants.

hybrid – an organism that has two different alleles for a particular trait.

hypothesis – a guess, or prediction, based on previous observations or prior knowledge that will be tested during an experiment.

igneous rocks – one of the three basic types of rock; forms when liquid magma cools and becomes solid.

independent variable – the variable that is changed or tested in an experiment; also called the manipulated variable.

indirect relationship – a relationship between variables whereby as one variable increases the other variable decreases.

inertia – the tendency of an object to resist a change in motion.

inference – a conclusion based on evidence, like data and observations.

inherited traits – characteristics of an organism that are controlled by genes that are passed from one generation to the next. Inherited traits can be used to classify organisms.

inner core – the layer of Earth believed to be made of extremely hot, dense, solid iron.

inquiry – the act of questioning or investigating.

instruments – technological tools, such as telescopes, thermometers, and microscopes, that help us collect data and observe our world.

International System (SI) of Measurement – an accepted standard of measurement that uses base units such as seconds, meters, and kilograms.

interpolation – estimating a value between two known values.

invertebrates – organisms that do not have a backbone.

jetty – a pier constructed of boulders and concrete that extends from land out into the ocean to protect the shore from storms and erosion.

joule – the unit of measurement for energy.

Jupiter – a large gas giant, outer planet, which is the fifth planet from the Sun.

kilogram – the SI base unit for measuring mass.

kinetic energy – a state of energy; energy found in moving objects.

kingdom – the largest and most general grouping of organisms within the biological classification system.

Kuiper belt – the region past the orbit of Neptune, which is composed of many icy bodies and the dwarf planet Pluto.

land breeze – the wind that occurs when air from over the land moves toward the sea.

lava – hot, molten rock on the surface of the Earth.

law of conservation of energy – a scientific law stating that energy cannot be created or destroyed by ordinary means, but it can be changed from one form to another.

law of conservation of matter – a scientific law stating that matter cannot be created or destroyed.

law of lateral continuity – a scientific law stating that layers of sedimentary rock extend laterally in all directions.

law of superposition – a scientific law stating that sedimentary layers are deposited in a time sequence; older rock layers are on the bottom and younger rock layers are on the top.

law of universal gravitation – a scientific principle by Isaac Newton that states that every object in the universe attracts every other object, and that this force is dependent upon the masses of the objects and the distance between them.

life cycle – the series of changes an organism goes through from birth to death.

line graph – a visual representation of continuous, quantitative data.

liquid – the phase of matter that has a definite volume, but will take the shape of whatever container it is in.

liter – the SI base unit for measuring the volume of a liquid.

lithification – the process in which sediments are compacted and cemented together, forming sedimentary rocks.

lithosphere – the layer of the Earth composed of the crust and the uppermost solid portion of the mantle.

longitudinal wave – a wave in which the matter in the medium moves back and forth in the same direction as the wave; also called a compressional wave.

lunar eclipse – an eclipse that occurs when the Moon is full and the Earth is between the Sun and the Moon, blocking the sunlight from illuminating the surface of the Moon.

luster – a glossy appearance; a physical property of metals.

luster – a property of a mineral that describes how light interacts with it; for example, waxy, oily, or metallic.

magma – hot molten rock material found deep below the Earth's surface.

malleable – the property of a substance that allows it to be shaped and formed by hammering and through pressure; a physical property of metals.

manipulated variable – the variable that is changed or tested in an experiment; also called the independent variable.

mantle – the thickest layer of the Earth consisting of hot, solid rock.

Mars – an inner, terrestrial planet, which is the fourth planet from the Sun.

mass – the amount of matter in an object.

matter – anything that takes up space and has mass.

mean – determined by dividing the sum of a set of data quantities by the number of quantities in the data set.

mechanical energy – the form of energy found in objects in motion.

mechanical waves – waves that require a medium, or substance, to travel through, such as ocean waves or seismic waves.

mechanical weathering – the disintegration of rocks into smaller particles due to interaction with water and wind; also called physical weathering.

median – the middle value in a set of numbers.

medium – the substance that carries mechanical waves.

meiosis – the process of cell division in which a cell divides twice, resulting in four sex cells, each with half the genetic material of the original cell.

melting point – the temperature at which a solid will turn into a liquid.

meniscus – the curved surface of the liquid as it clings to the sides of the container.

Mercury – the inner, terrestrial planet which is closest to the Sun.

metal – elements that are good conductors, are malleable and ductile, and have a glossy appearance.

metamorphic rocks – one of the three basic types of rock; form deep below the surface of the Earth where heat and pressure change the preexisting solid rocks.

metamorphism – physical and chemical changes in the rock caused by intense heat and pressure.

meteorologists – scientists who study weather, observe weather changes, and make predictions based on previously observed weather conditions.

meter – the SI base unit for measuring length.

mineral – naturally occurring inorganic substances that have a definite chemical composition.

mitochondria – sometimes called the "powerhouse of the cell," because through the process of respiration, they provide the cell with energy.

mitosis – the process of cell division in which a cell divides into two new body cells; these new cells are identical to the original cell.

mixture – a combination of two or more substances in varying amounts that are physically combined.

mode – the most frequent value that appears in a set of numbers.

Moh's hardness scale – a scale used to determine a mineral's resistance to being scratched; can be used to help identify a mineral.
molecules – two or more atoms joined together forming a compound.
multicellular – organisms made of many cells.
mutualism – a symbiotic relationship in which both organisms benefit from the relationship.

natural selection – the survival and reproductive success of organisms that are better adapted to an environment because of beneficial traits.
neap tides – the lowest tides, occurring twice per month, when the Sun and Moon are at a 90-degree angle relative to the Earth.
Neptune – a large gas giant, outer planet, which is the eighth planet from the Sun.
neutrons – particles with no electric charge found in the nucleus of atoms.
Newton's laws of motion – three scientific laws stated by Isaac Newton that explain the motion of all objects in the universe.
niche – where an organism lives and its role within the ecosystem.
noble gases – elements found in the last column, or group, on the Periodic Table of Elements, that are usually in the gas state and normally do not react with other elements.
nonmetal – elements that are poor conductors, have a dull appearance, and are neither malleable nor ductile.
nonrenewable resources – resources such as mineral deposits, coal, oil, and natural gas, that cannot be replenished because they take millions of years to form naturally.
nuclear energy – the form of energy stored in the nucleus of atoms.
nucleus (of a cell) – controls all the cell's activities and contains the genetic information for the organism.
nucleus (of an atom) – the center of an atom, which consists of protons and neutrons.

observations – noticing an event or specimen in a careful and thorough way and recording what is observed in a detailed manner.
ocean – the largest biome covering the majority of the Earth's surface and containing a variety of marine organisms.
ocean currents – the circulation of ocean waters, mainly due to the uneven heating of the Earth's surface.
omnivore – consumer that feeds on both plants and animals.
opaque – objects that scatter and/or absorb light.

order – a grouping between class and family within the biological classification system (kingdom, phylum, class, order, family, genus, species), where kingdoms are broad, general groupings of organisms and species are more specific groupings.
organ – different tissue layers that join together to perform specialized functions for an organism, such as the heart and lungs.
organ system – organs working together to carry out major bodily functions, such as digestion and circulation.
organisms – living things.
outer core – the layer of Earth believed to be made of dense, liquid iron and nickel.
ozone layer –the layer of the Earth's atmosphere found six to thirty miles above the Earth's surface that absorbs ultraviolet radiation from the Sun.

parasite – the organism that benefits within a parasitic relationship.
parasitism – a symbiotic relationship in which one organism benefits while the other organism is harmed.
Periodic Table of Elements – an organized table that contains information about each element and the relationships and interactions between them.
periods – the rows on the Periodic Table of Elements.
permafrost – permanently frozen soil found in the tundra biome.
phases of the Moon – the changing appearance of the Moon at various times of the month as different portions of the Moon appear to be illuminated.
phenomena – observable natural events.
phenotype – the physical appearance of a particular trait.
photosynthesis – a process in plant cells where chloroplasts convert light energy from the Sun into chemical energy stored as sugar.
phylum – a grouping between kingdom and class within the biological classification system (kingdom, phylum, class, order, family, genus, species) where kingdoms are broad, general groupings of organisms and species are more specific groupings.
physical change – a change in matter in which the composition of the substance stays the same but the shape, phase, or location may be altered.
physical properties – properties of a substance that can be measured and observed and that do not change depending on the size of the sample of matter.
physical weathering – the disintegration of rocks into smaller particles, due to interaction with water and wind; also called mechanical weathering.

phytoplankton – microscopic free-floating plants found in the ocean biome.

plants – one of the six kingdoms of living things; multicellular eukaryotes that are autotrophs, such as mosses, ferns, trees, and flowering plants.

plate tectonics – the theory stating that the lithosphere is broken up into plates that float independently over the asthenosphere at a rate of centimeters per year.

population – a group of organisms of the same species that are found together in a given place and time.

potential energy – a state of energy; energy stored in objects.

prairies – the biome where the main vegetation is a variety of grasses; sometimes called savannas or grasslands.

precipitation – the process that occurs when condensed water particles in clouds become too large to remain suspended and fall to the surface.

precision – the ability of a measurement to be consistently reproduced.

predation – the interaction between organisms whereby one animal hunts and consumes other animals.

predator – an animal that lives by hunting and feeding upon another organism.

prey – an animal that a predator hunts and feeds upon.

primary consumers – the herbivores in a food chain; the organisms that consume the producers.

primary producer – an organism that makes its own food from nonliving sources and that makes up the largest amount of biomass in an ecosystem.

producer – an organism that makes its own food from nonliving sources.

product – a substance that is produced or present after a chemical reaction or chemical change.

prokaryote – an organism composed of a cell or cells that do not contain a nucleus.

protists – one of the six kingdoms of living things; mostly unicellular eukaryotes, such as euglena and amoebas.

proton – positively charged particles found in the nucleus of atoms.

Punnett square – a chart used to determine the chance, or probability, of a particular trait appearing in the potential offspring of two organisms.

purebred – an organism that has two of the same alleles for a particular trait.

qualitative – a descriptive observation rather than a measurement.
quantitative – an observation in the form of a measurement.

radial symmetry – symmetry found in organisms like starfish and jellyfish; symmetry around a central axis; organisms may resemble a pie shape.
radiation – the transfer of heat through empty space.
reactants – a substance that is present at the beginning of a chemical reaction or chemical change.
recessive – a weaker trait or gene; a trait that can be masked by dominant traits.
reflection – when light strikes a surface at an angle, then bounces off at an equal angle.
refraction – the bending of light as it transmits through an object, due to the change in the speed of light within various materials.
renewable resources – resources such as water, air, soil, and solar energy, which are replenished naturally over the course of time.
repeated trials – replicating an experiment a few times to reduce the effects of errors and unidentified variables on data and results.
respiration – the process by which an organism takes in oxygen and releases carbon dioxide. In cells, mitochondria use oxygen and sugar to produce energy, carbon dioxide, and water.
responding variable – the variable that changes in response to the independent variable; also called the dependent variable.
revolution – the movement, or orbiting, of one object around another.
Richter scale – a scale used to measure the magnitude of an earthquake, using the largest seismic wave recorded by a seismograph during the earthquake.
rock – naturally occurring substances made of one or more minerals.
rock cycle – the process in which rocks are formed, changed, melted, and reformed as igneous, sedimentary, and metamorphic rocks.
rotation – the spinning of an object on an imaginary axis.
runoff – the movement of water across the land until it collects in lakes, oceans, soil, and rocks underground.

Saturn – a large gas giant, outer planet, which is the sixth planet from the Sun and has an elaborate ring system.
savannas – the biome where the main vegetation is a variety of grasses; sometimes called grasslands or prairies.
scattering – when light bounces off an object in various directions after striking it.

scavenger – animals that eats dead animals, or carrion.
sea breeze – the wind that occurs when air from over the sea moves toward the land.
sea floor spreading – a theory that suggests that oceanic crust is created through volcanic activity along ocean ridge systems.
second – the SI base unit for measuring time.
secondary consumers – the omnivores and carnivores that consume the small herbivores in a food chain.
sedimentary rocks – one of the three basic types of rock; form mostly when sediments are compacted and cemented together; may contain fossils.
sediments – small particles of rock such as gravel, sand, silt, and clay, produced by the physical and chemical weathering of rocks.
seismic waves – the movement of energy that is released during an earthquake.
seismographs – an instrument used to measure seismic waves.
sexual reproduction – a form of reproduction in which sex cells combine to create a new organism that has a mixture of traits from both parents.
soil – weathered rocks and decomposed organic material from dead plants, animals, and bacteria, found in a very thin layer on the Earth's land surfaces.
solar eclipse – an eclipse that occurs when the Moon is directly between the Sun and Earth, and the Moon blocks the sunlight.
solar system – consists of the Sun and the revolving celestial bodies, such as planets, dwarf planets, asteroids, and meteoroids.
solid – the phase of matter that has a definite shape and a definite volume.
soluble – a material that can be dissolved in a solution.
solute – the substance that is dissolved in a solution.
solution – a type of mixture in which one substance in dissolved in another.
solvent – in a solution, the substance that the solute is dissolved in.
sound – a type of mechanical energy released when an object vibrates and sound waves are generated that can move, or travel, through matter.
species – the most specific grouping of organisms within the biological classification system (kingdom, phylum, class, order, family, genus, species).
spring tides – the highest tides, which occur twice per month during the new and full Moon, when the Earth, Moon, and Sun are aligned.
streak – the color of the powder left behind after being rubbed across a hard surface; can be used to help identify a mineral.

subduct – the sinking of crust into the upper mantle, due to the collision or converging of plates and the density of the rock in the crust.
succession – the fairly predictable series of communities that will replace one another as an ecosystem develops over time.
supernova – the explosion of a star as it dies.
symbiosis – a relationship in which two organisms live in close vicinity to one another and at least one organism benefits from the relationship.

telescope – an instrument that improves our ability to observe stars and our solar system by magnifying our view of these celestial objects.
temperate deciduous forest – the biome that is characterized by having trees that lose their leaves each year.
tertiary consumers - the animals that feed on the secondary consumers in a food chain.
thermometer – an instrument used to measure temperature.
tides – the regular and predictable motion of ocean water, caused by the gravitational attraction between the Earth, Moon, and Sun.
tissue – similar cells that join together to perform a special function for an organism.
transform plate boundaries – the location where two lithospheric plates slide past each other.
translucent – objects that transmit and scatter light.
transmission – the passing of light through an object.
transparent – objects that transmit light.
transpiration – the evaporation of water directly from the leaves of plants.
transverse wave – a wave in which the matter in the medium moves up and down, or perpendicular, to the direction of the wave.
triple beam balance – an instrument used to determine the mass of an object.
trophic level – the levels of consumption in a food chain.
tropical rain forest – the biome that occurs in areas with high rainfall and high temperatures, with a layered vegetation and great biodiversity.
trough – the bottommost part of a wave.
tundra – the biome found in the very cold, windy Arctic regions, with very little precipitation and permanently frozen soil.

unbalanced forces – forces that will cause a change in the speed or direction of an object's motion.
unicellular – organisms made of only one cell, such as bacteria.

Uranus – a large gas giant, outer planet, which is the seventh planet from the Sun.

vacuole – storage areas in a cell that store food, water, and waste.
variable – any part, or component, of an experiment that can change or be changed.
Venus – an inner, terrestrial planet, which is the second planet from the Sun.
vertebrate – an organism that has a backbone.
volume – the amount of space an object takes up.

warm front – the boundary where a warm air mass pushes against and over a cold air mass.
water cycle – the circulation or movement of water through an ecosystem.
water displacement method – a method of determining the volume of an irregular solid by measuring the volume of the liquid displaced.
wavelength – the distance between any two adjacent corresponding locations of a wave.
waves – traveling disturbances that carry energy.
weather – the state of the atmosphere at a given place and time.
weather map – a map used to represent and predict weather conditions.
weathering – the process of rock being broken down into smaller pieces.
wetlands – a water biome consisting of marshes, swamps, and bogs that form a transitional area between dry land and bodies of water, and that provide a habitat for many bird species.

INDEX